THE VITAL
MACHINE

THE VITAL MACHINE

A Study of Technology and Organic Life

DAVID F. CHANNELL

New York   Oxford
OXFORD UNIVERSITY PRESS
1991

Oxford University Press

Oxford   New York   Toronto
Delhi   Bombay   Calcutta   Madras   Karachi
Petaling Jaya   Singapore   Hong Kong   Tokyo
Nairobi   Dar es Salaam   Cape Town
Melbourne   Auckland

and associated companies in
Berlin   Ibadan

Published by Oxford University Press, Inc.
200 Madison Avenue, New York, NY 10016

Oxford is a registered trademark of Oxford University Press

Library of Congress Cataloging-in-Publication Data
Channell, Daivd F., 1945–
The vital machine : a study of technology and organic life /
by David F. Channell.
p.   cm.
ISBN 0-19-506040-7
1. Technology—Philosophy.   I. Title.
T14.C464   1991
601—dc20   90-20273   CIP

2 4 6 8 8 9 7 5 3 1

Printed in the United States of America
on acid-free paper

To Carolyn Eilmann Channell

Acknowledgments

I owe many people a debt of gratitude for their assistance, both directly and indirectly, in the completion of this book. Melvin Kranzberg and Robert Schofield introduced me to the history of science and technology as a scholarly discipline. Reese Jenkins stimulated my interest in the relationship between technology and the organic. Edwin Layton made me think seriously about the complicated relationships that exist between science and technology. He also read a draft of the entire manuscript and gave me many helpful suggestions.

Much of the work on this book began during a year that I spent as a fellow at the National Humanities Institute at the University of Chicago. During that year Stephen Toulmin, W. David Lewis, Frederick Ferré, James Gustafson, and Hans Jonas helped me to formulate many of the ideas in this book. Subsequent grants from the National Science Foundation (SES-8015514), the National Endowment for the Humanities (EH-20029-80-2058), and the University of Texas at Dallas allowed me to refine many of my ideas. My colleagues and students at the University of Texas at Dallas provided a stimulating environment, which allowed many of my ideas to come to fruition.

I could not have completed this work without the assistance of the following institutions and their libraries: the University of Texas at Dallas, Case Western Reserve University, the University of Chicago, Southern Methodist University, the British Library, the National Library of Scotland, the University of Edinburgh, Glasgow University, and the University of Cambridge.

The editors and staff at Oxford University Press have been very professional and supportive, especially Senior Editor Cynthia Read and Associate Editor Ruth Sandweiss. My wife Carolyn took valuable time from her own work to provide me with much needed advice and suggestions on the content and style of the book.

Contents

Illustrations follow page 84

THE VITAL
MACHINE

1

Introduction: The Mechanical and the Organic

And was Jerusalem builded here
Among these dark Satanic Mills?

WILLIAM BLAKE, *Milton*

The Problem

One of the most important issues facing us as we move toward the twenty-first century is a new relationship between technology and organic life. Recent developments in the areas of genetic engineering, biomedical engineering, and artificial intelligence are raising profound questions about the definition of life and what it is to be human. Genetic engineering gives us the ability to create new forms of life. Human genes can be inserted into animals, and vice versa, leading to transgenic species that are neither animal nor human. Already some one thousand transgenic varieties of mice exist, including one with a human growth gene.[1] Biomedical engineering through the creation of kidney machines, ventilators, pacemakers, artificial hearts, incubators, and in vitro fertilization is raising new questions about when life begins and ends.[2] Since machines can maintain heartbeat and respiration, sometimes indefinitely, many states have "redefined" death as the cessation of brain activity. But what will happen if and when some form of brain activity can be artificially maintained? Already a cat's brain has been kept "alive" for a day outside its body. Does this mean the cat is still alive or does it even make sense to call such a disembodied brain a cat? Artificial intelligence is developing computer programs that can reason out problems, make decisions, perceive, interact with their environments, and learn.[3] All of this raises questions about the uniqueness of human intelligence and about whether some machines should be considered conscious beings. Already a computer has been able to beat a grand master at chess and passengers are flying in airplanes in which computers watch over the pilots and keep them from making maneuvers that might endanger the plane.

3

Why do we find these new developments in genetic engineering, biomedical engineering, and artificial intelligence so difficult to understand and so troubling? Do these new developments have anything in common? I think an answer to the second question will help us begin to address the first. Although research in genetics and medicine seems to have little in common with research in computer engineering, developments in all three areas force us into situations where we cannot distinguish between what is technological and what is organic. Each area contains one term, such as *genetic, biomedical,* or *intelligence,* which we associate with biology and the organic. Then these terms are juxtaposed with other terms, such as *engineering* or *artificial,* which we associate with technology and the mechanical. That is, these recent developments in each area exist in a netherland between the technological and the organic and fall outside our traditional systems of classification.

The anthropologist Victor Turner has used the term *liminal* from the Latin term for threshold, to refer to such in-between states.[4] Traditionally human culture has been intolerant of ambiguity. We tend to find liminality strange and uncomfortable. One can be in a space or outside a space, but we find it unsettling if a person is on a threshold or standing in a doorway. Another anthropologist, Mary Douglas, has noted that transitional states, things that do not fit into our accepted way of classifying the world, are seen as evil or dangerous.[5] She argues that historically our concept of dirt or pollution is more concerned with a violation of our classification system, with things being out of place, than with hygiene or disease-causing agents. For example, a room can be considered dirty if things are out of place—clothes on the floor, shoes on the bed—even if it is hygienic. Foods forbidden in biblical dietary laws also fell outside traditional classification schemes. Camels and pigs were considered unclean because they did not both chew their cuds and have cloven hoofs as did other animals classified as livestock.

The liminal or transitional nature of genetic engineering, biomedical engineering, and artificial intelligence raises questions about how we classify things as technological or organic. To understand these developments we must re-examine the relationship between technology and the organic. In conducting such a study we may be led to discover a new way to understand the world so that we can intelligently and responsibly deal with these new developments. They will no longer be liminal because we will in a sense have "moved the door." Mary Douglas believes that when we discover something that falls outside our traditional schemes of classification "we can deliberately confront the anomaly and try to create a new pattern of reality in which it has a place."[6] In a similar way Victor Turner sees liminal situations "as the settings in which new models, symbols, paradigms, etc., arise—as the seedbeds of cultural creativity in fact."[7]

Since at least the time of the Renaissance, it has been a common practice to divide the world into two distinct and separate categories—the technological (or mechanical) and the organic (or natural). If we examine the differences between technology and the organic we will discover that they exist on many levels and are much more complex than we might originally think. Because of all the reports we see about the damaging effects technology has had on nature, we tend to see a fundamental and irreconcilable opposition between the two. But the distinction between technology and the organic is not as simple as suggested by the common image of machines encroaching on nature. Technology has had detrimental effects on organic life; air pollution, acid rain and toxic waste dumps are obvious examples.[8] Natural events, such as hurricanes, tornadoes, volcanic eruptions, floods, and climatic changes, also regularly bring destruction to all levels of life. While chainsaws have deforested large areas of the wilderness and strip mining has changed the shape of the landscape, the eruption of Mount St. Helens and the forest fires started by lightning have added to the devastation of significant wilderness areas. Technology has certainly been responsible for the extinction of species of plants and animals, but the fossil record indicates that large-scale extinctions were taking place long before the evolution of humans and their technology. A meteor impact was probably responsible for the worldwide pollution of the atmosphere that led to the extinction of the dinosaur. In the early history of Earth, oxygen was a poisonous toxic gas which had deadly consequences for most organisms that existed at that time. So, on the one hand, epidemics like the plague and AIDS are products of nature as are many of the world's carcinogens. On the other hand, many parts of our "natural" environment, such as domesticated animals, agricultural crops, and hybrid plants, are the result of a long history of human intervention and design. Human beings are frail creatures who would have a difficult time surviving in a completely "natural" world. Throughout history we have depended on technology for shelter from the elements, for preparation and storage of food, for sanitation, for communication, and for transportation.

An Opposition of Values

Many recent social critics see the conflict between technology and the organic existing on another level.[9] They see this difference as an opposition between the values associated with technology and the organic. For them, the machine and nature represent different sets of ideas and feelings about the world and our relationship to it.

The idea that technology and the organic represent different value systems can be traced to the intellectual, literary, and artistic movement of the late eighteenth and early nineteenth centuries that has been labeled Romanticism. Many of today's attitudes toward technology and the organic have their roots, either directly or indirectly, in the conclusions of the Romantics.[10] Although the term *Romanticism* is most often applied to a literary movement, it encompassed the entire realm of thought.[11] According to Alfred North Whitehead, "the romantic revival was a protest on behalf of the organic view of nature, and also a protest against the exclusion of value from the essence of matter of fact."[12] It is a common assumption that the main focus of these protests was science and technology. But the Romantics were not necessarily opposed to all forms of science and technology. While some writers, such as William Blake, were opposed to technology, others maintained an active interest in scientific and technological matters. For example, Goethe conducted important research on biology and the theory of light;

F. W. J. von Schelling was elected secretary of the Academy of Sciences in Munich; Novalis was an amateur scientist; Percy Bysshe Shelley believed that science and poetry could bring about a new future; and Samuel Taylor Coleridge was friends with many scientists of his day, lectured at the Royal Institution, and wrote a book on biology.[13] Henry David Thoreau called himself an engineer even while living at Walden Pond.[14] In the preface to his *Lyrical Ballads,* William Wordsworth argued that the poet "will be ready to follow the steps of the man of science, not only in those general indirect effects, but will be at his side, carrying sensation into the midst of the objects of the science itself. The remotest discoveries of the chemist, the botanist, or the mineralogist will be as proper objects of the poet's art as any upon which it can be employed."[15]

The protests of the Romantics were aimed at the mechanical philosophy that lay at the foundation of much of eighteenth-century science and technology.[16] For many Romantics, the distinction between the mechanical and the organic reflected different capacities of the human mind. Enlightenment philosophers, such as John Locke, David Hume, and David Hartley, saw the mind functioning according to mechanical laws.[17] For them the "ideas" that existed in the mind were simply the images of sense perceptions of the external world. In this view, the creative activity of the mind was seen as a mechanical process in which ideas were broken down into their component parts and rearranged into new wholes, analogous to a mechanic using pieces of several machines in order to make a new invention. The Romantics, however, argued that the mind functioned more like a plant than a machine.[18] Although some faculties of the mind, such as memory, might function according to mechanical laws, writers like Coleridge argued that the higher faculties, such as the imagination, were "essentially *vital.*"[19] While the mechanical mode of the mind simply transposed data and facts, the vital mode was "synthetic," and "permeative."[20] Coleridge summarized the differences between the two modes of thought: "The mechanic system . . . knows only of distance and nearness . . . in short, the relations of unproductive particles to each other; so that in every instance the result is the exact sum of the component qualities, as in arithmetical addition. . . . In Life . . . the two component counter-powers actually interpenetrate each other, and generate a higher third including both the former."[21]

The Romantics' protest on behalf of the organic was not directed at science and technology but at the use of the machine as an *image* of thought and culture.[22] In his essay, "Sign of the Times" (1829), Thomas Carlyle analyzed the role of the machine as a sign or symbol of the emerging modern world. He explains, "Were we required to characterize this age of ours by any single epithet, we should be tempted to call it, not an Heroical, Devotional, Philosophical, or Moral Age, but above all others, the Mechanical Age. It is the Age of Machinery."[23] Carlyle notes that the term *Age of Machinery* can be used in two different ways. In its most obvious sense, the term refers to the material effects brought about by technological development. Here the results of mechanism have been both negative and positive. On the one hand, "the living artisan is driven from his workshop, to make room for a speedier, inanimate one," while on the other hand, "what wonderful accessions have thus been made . . . to the physical power of mankind; how much better fed, clothed, lodged and, in all outward respects, accommodated men now are.''[24]

But Carlyle's concern is with the more inward sense of the Mechanical Age, with "how the mechanical genius of our time has diffused itself into quite other provinces. Not the external and physical alone is now managed by machinery, but the internal and spiritual also.''[25] This internal sense of the mechanical represents a set of values that has affected education, religion, philosophy, science, art, literature, and music.[26] In each of these areas mechanism leads to an emphasis on "combinations and arrangements" rather than on "internal perfection.''[27] Carlyle believed that the extension of mechanical ideas to the philosophy of the mind led to an emphasis on the origins of consciousness but neglected the "grand secrets" of free will, time, space, God, and the universe.

Carlyle gave the term *dynamical* to the set of values neglected by the mechanical. The dynamical treated "the primary, unmodified forces and energies of man, and the mysterious springs of Love, and Fear, and Wonder, of Enthusiasm, Poetry, Religion.''[28] The dynamical was distinguished from the mechanical by its vital character. According to Carlyle, elements considered dynamical "rose up, as it were, by spontaneous growth, in the free soil and sunshine of Nature.''[29]

The Machine and the Organic as Symbols

In "Sign of the Times," the machine and the organic became more than ordinary metaphors. Carlyle said, "Considered merely as a metaphor, all this is well enough; but here, as in so many other cases, the 'foam hardens itself into a shell,' and the shadow we have wantonly evoked stands terrible before us and will not depart at our bidding.''[30] For the Romantics, the machine and the organic became symbols—what philosopher Stephen Pepper has called root metaphors.[31] A root metaphor is some commonsense object or fact that is used as a basic analogy to help people understand the world. In his book *The Machine in the Garden,* Leo Marx refers to the machine and the garden as "cultural symbols," which he defines as images that convey "a special meaning (thought and feeling) to a large number of those who share the culture.''[32]

In this book I wish to argue that the basis of the current tension between technology and organic life does not arise as a conflict between machines and nature. Instead, it must be understood in terms of a tension between the machine and the organic as root metaphors or cultural symbols. Recent theories of symbolism can provide some insights into how machines and organic life "can create, revise, transform, and re-create wholly fresh products, systems, and even worlds of meaning.''[33] In a symbol we recognize that a concrete object is standing for something else, usually something abstract and general. One of the most comprehensive theories of symbolism was proposed by Ernst Cassirer in such works as *The Philosophy of Symbolic Forms,* and *An Essay on Man,* in which he argues that symbols are a necessary part of human thought. It is through the use of symbols that the mind understands what is called reality. For Cassirer, "all mental processes fail to grasp reality itself, and in order to represent it, to hold it at all, they are driven to the use of symbols.''[34] Rather than defining humans as *animals rationale,* he finds it better to define them as *animals symbolicum.*[35] Humans do not live in a world of physical objects but in a symbolic world. According to Cassirer:

> No longer can man confront reality immediately; he cannot see it, as it were, face to face. Physical reality seems to recede in proportion as man's symbolic activity advances. Instead of dealing with the things themselves man is in a sense constantly conversing with himself. He has so enveloped himself in linguistic forms, in artistic images, in mythical symbols or religious rites that he cannot see or know anything except by the interposition of this artificial medium.[36]

Cassirer's theory of symbolism was further articulated by his student Susanne Langer in her book *Philosophy in a New Key*. Like Cassirer she argues that what ''sets man so far above other animals . . . is the power of using symbols.''[37] For Langer the use of symbols has to be distinguished from the use of signs.[38] A sign takes the form of a conditioned response. In Pavlov's famous experiments, the ringing of a bell became a proxy for some other object, and it led to some action as if the object were present. Although a symbol can have the same referent as a sign, it has a different function. Langer argues, ''Symbols are not proxy for their objects, but are *vehicles for the conception of objects . . . it is the conceptions, not the things, that symbols directly 'mean.' ''*[39]

One of the most important aspects of symbols is their ability to transform our perceptions about the world. It is through certain symbols that we project a sense of values onto the world.[40] Anthropologist Clifford Geertz has argued that synthesizing symbols ''relate an ontology and a cosmology to an aesthetics and a morality: their peculiar power comes from their presumed ability to identify fact with value at the most fundamental level, to give to what is otherwise merely actual, a comprehensive normative import.''[41] Geertz's synthesizing symbols, Leo Marx's cultural symbols, or Stephen Pepper's root metaphors delimit what has been called a world view or world hypothesis. Geertz defines a world view as the cognitive, existential aspects of a culture. ''[A culture's] world view is their picture of the way things in sheer actuality are, their concept of nature, of self, of society. It contains their most comprehensive ideas of order.''[42] Since the time of Immanuel Kant, philosophers such as Cassirer, E. H. Gombrich, Ludwig Wittgenstein, Jean Piaget, Jerome Bruner and Nelson Goodman have argued that a world view is, in fact, a mental construction that is actively imposed onto the world.[43] According to Nelson Goodman, ''The many stuffs—matter, energy, waves, phenomena—that worlds are made of are made along with the worlds.''[44]

Mechanical and Organic World Views

Although a culture usually is defined by a single world view it is possible for a culture to have two different yet coexisting sets of beliefs and attitudes which result in two coexisting world views, each represented by its own cultural symbol or root metaphor. By focusing on the machine and the organic as symbols which represent two different world views, I will argue that the perceived tension between technology and organic life has been a tension between opposing world views. Each world view determines, among other things, the models that people use to understand technological developments and organic processes. Throughout much of Western history a tension has existed between two world views—one that can be labeled mechanical and the other organic. In some form each of those views has been present throughout the past but each view has not remained static.[45] In the course of history each world view has undergone development or change, in many cases because of interactions with the other view. During certain periods, one or the other view took its turn at dominance and each world view established its own ideas concerning the relationship between technology and organic life. The same technological developments or organic processes take on completely different values depending on the world view from which they are seen.

In the first part of the book I will analyze the historical development of the mechanical world view, which came to use the symbol of a machine, especially a clock, to understand the world. According to this view, the world is made up of unchangeable and indistinguishable pieces of matter whose only difference is their position and motion through space. In this view phenomena can best be understood by reducing them to their simplest constituent parts. For the mechanist a complete understanding of any complex phenomenon can be gained from a study of its separate parts. That is, the whole is equal to the sum of its parts. In such a world view, there is no conflict between actual machines and organic processes since both technology and life will be thought to be based on mechanical principles. Here I will focus on how the symbol of the machine transformed life into what I call mechanical organisms. From the Renaissance to the twentieth century, medicine has come to use technological models to explain various organic processes, beginning with the circulation of the blood, digestion, respiration, and eventually even mental activity. Belief in these models led to attempts to actually recreate life through technological means.

In the second part of the book I will analyze the development of an opposing organic world view, which used the symbol of an organism, such as the body or a plant, to understand the world. Rather than focusing on isolated parts, the organic view emphasizes the organization or relationship between parts. For the organicist the organization of parts into a whole results in qualitatively new phenomena such as a vital spirit, principle, or force. That is, the whole is more than the sum of its parts and the nature of the parts is determined by the whole.

In such a world view there is also no conflict between machines and organic processes since both will be thought to arise from some vital organization. This part of the book will focus on how the symbol of the plant transformed technology into what I call organic machines. From Aristotle to the Renaissance alchemists and magicians, and later to Karl Marx, the symbol of the organism has been used to explain technological developments.

Throughout the course of history the tension between the two world views has resulted in both sides' making concessions and incorporating some elements of the opposing view into their own. During the twentieth century the distinctions between the two world views has faded; a new world view has begun to emerge. In the third part of the book I will analyze this development: a new bionic world view that transcends the earlier thinking. Unlike the reductive approach of the mechanical view or the holistic approach of the organic view, the bionic world view is consciously dualistic in its understanding of the world. This world view is based on a new symbol that I call the vital machine, which emphasizes the role of interactive processes or dualistic systems in understanding the world. This section of the book will focus on how the symbol of the vital machine provides new insights into the relationship between technology and life in light of recent developments in artificial intelligence, genetic engineering, and biomedical engineering. I will show that the emergence of the vital machine raises a wide range of new ethical problems, and, if we are to solve these problems, traditional human values will have to undergo a significant transformation.

2

The Mechanical World View: The Clockwork Universe

> He that knows the structure and other mechanical affections of a
> watch, will be able by them to explicate the phaenomena of it,
> without supposing, that it has a soul or life to be the internal
> principle of its motions or operations.
>
> ROBERT BOYLE, *The Vulgarly Received Notion of Nature*

The so-called term *mechanical philosophy* emerged from the extraordinary period
during the sixteenth and seventeenth centuries known as the Scientific Revolution
and has come to be associated with a diverse range of approaches to the world,
including atomistic, corpuscularian, materialistic, hypothetical, and experimental. Over time it has undergone significant changes.[1] But in all of its forms,
mechanical philosophy has represented a reductionist approach to understanding
the world—an approach that views the world as functioning like a machine,
particularly a clock. Some of the elements that contributed to a mechanical world
view appeared as far back as classical antiquity. Three important sets of ideas—
atomistic, mathematical, and mechanistic explanations of natural phenomena—
can be found in ancient Greek civilization.[2]

One of the problems that faced Greek philosophers was an explanation of the
nature of the apparent change they saw in the world. Their goal was to find some
stable, unifying pattern behind the change. One attempt to discover such a
principle developed during the sixth and fifth centuries B.C., when Pythagoras put
forward the theory that the world was made up of numbers. That is, numbers had
physical properties; for example, three was triangular and four was square so that
the physical world could be constructed from these numerical units. Any change
that seemed to take place in the world was an illusion since behind it there existed
the unchanging world of numbers. But, as Aristotle noted, it was hard to conceive
of a world made up of actual numbers.[3]

A more acceptable theory was proposed by Leucippus and developed by
Democritus. They postulated that the world consisted of small, indivisible, ever
moving material units—atoms. These atoms, only distantly related to modern day
atoms, could be considered as the physical counterpart of Pythagoras's numerical

11

units. Although all atoms were composed of the same material, they could combine into different shapes and patterns. Therefore, the changing appearance of the world could be explained by the unchanging properties of atoms—size, shape, and motion. But since atoms were so small, they could not be experienced through the senses. The only way to gain any knowledge of them was through the rational powers of the mind. In fact, at some point Democritus implied that atoms themselves were not physically real but were simply theoretical entities that allowed him to understand the world.

Although atomism did not become the dominant theory of nature during the Classical period, many of its ideas were incorporated into later intellectual traditions. Through it the normal world that we experience with our senses was reduced to a world of ideal sizes, shapes, and motions.

In a parallel manner, Plato's theory of knowledge reduced the world of the senses to an ideal world of mathematical forms. Like other Greek philosophers, Plato was concerned with the problem of change. How could we have any true and absolute knowledge of the world if it was undergoing continual change? His answer came in his analogy of the cave.[4] Plato asked us to picture people in an underground cave who were chained in such a way that they could see only the back wall of the cave. The light from outside could shine through the entrance onto the back wall. Outside the cave, other people were carrying various objects past the entrance of the cave. The chained people inside the cave never see the objects themselves; they see only the objects' shadows on the back wall of the cave. From these shadows they must try to infer the nature of the real objects. This task will not always be easy; the same object, if carried at different angles to the opening of the cave, will cast different shadows. Only the most astute observer in the cave will discover that several different shadows might have a common source.

For Plato, the cave was analogous to the world. The world that we experience through our senses is like the shadows on the back of the cave—constantly changing and never the same. But the "real" world is outside the cave; it is these "real" objects, what Plato calls the forms, that lie behind the shadows. For example, behind every horse that we see there is the ultimate form which is the very essence of "horseness." Each individual horse is only an imperfect copy, a shadow, of the ultimate form. In Plato's view it is impossible to have true knowledge about the world around us, since this world is only a shadow world anyway. The real world about which we should be concerned is the world of forms. The closest things to the forms were pure geometrical figures. We gain knowledge of this world not through the senses but through the rational powers of the mind. Therefore, Plato's philosophy reduces the world of our sense experience to a world of mathematical forms.

Greek astronomers paved the way for mechanistic models of the universe.[5] Since they assumed that celestial bodies, being perfect, moved in perfect circular paths, it became the goal of Greek astronomy to reduce the observed irregular motion of the planets around the earth to some form of uniform circular motion. One of the most successful solutions to the problem was put forward by Plato's student Eudoxus, who devised a system of concentric spheres centered on the earth. Each planet was connected to a sphere which in turn was connected to several other concentric spheres, each rotating around a different axis. By using up to twenty-seven spheres, Greek astronomers could reduce the complex pattern of planetary motion to the circular motion of the spheres. Later Apollonius and Hipparchus devised a model, subsequently adopted by the famous astronomer Ptolemy, in which a planet moved on a small circle, or epicycle, whose center moved on a larger circle called a deferent. As historian Otto Mayr has argued, these geometric models served as the basis for actual mechanical models such as the famous planetarium of Archimedes and also as the forerunners of the medieval astrolabe in which a flat disk, representing the stars, and movable grids for the planets, could be used to calculate the position of the heavens.[6]

The success of these geometric and mechanical models led many people to the conclusion that the universe was, in fact, a mechanical system.[7] For example, Aristotle argued that the universe was composed of fifty-five concentric crystalline spheres. A number of the spheres added by Aristotle functioned as mechanical linkages and "idle wheels" so that the motion of the outer sphere of the stars could drive all of the planets. Such beliefs in the reality of crystalline spheres continued, in varying degrees, until the Danish astronomer Tycho Brahe observed a comet in 1577 whose orbit would have had to penetrate a number of the spheres.

It cannot be argued that the Greeks solely established a mechanical world view. In fact, as we shall see, the foundation of the organic world view can also be found in classical Greece, sometimes in the same theories that we have just described. But the mechanical philosophers of the Scientific Revolution were greatly influenced by Greek atomism, Platonic forms, and mechanistic models.

The Mechanical Clock

Contrary to the common impression that the Middle Ages was a dark and dull era of Western history, the medieval period was one of the most fruitful in terms of advances in technology. Some historians, such as Lynn White, Jean Gimpel and Carlo Cipolla, give us a picture of the Middle Ages as containing the roots of our modern industrial society.[8] Wind and water power were harnessed, new navigational techniques were developed, and both cannons and movable type printing were invented. But one of the most significant inventions of the Middle Ages was the mechanical clock, which would make an important contribution to mechanical philosophy.[9]

Going back at least to early Egypt, we can trace an interest in timekeeping devices such as the sundial and the water clock. The ability to calculate the position of the planets, particularly the sun and the moon, was important in an era of widespread belief in astrology and of religious festivals and ceremonies determined by the position of the planets. By the Middle Ages the demand for an accurate timetelling device had increased. The monasteries, especially those following the rules of St. Benedict, needed to know the seven periods of prayer, and, in the towns, merchants and leaders needed to know the times to open and close the city gates, begin the markets, and enforce the curfew. Both sundials and water clocks had limitations; sundials could not be used at night or during cloudy, overcast days that were common in northern Europe, while water clocks, which relied on the changing level of water in a vessel, could freeze during the winter or on cold nights. Unfortunately, the name and location of the inventor of the first mechanical clock remain unknown, but the evidence suggests that by 1300 the basic weight-driven clock mechanism had been invented.[10] By the fourteenth century there are specific references to mechanical clocks, including the tower clock in Norwich Cathedral built by Roger Stokes, an astronomical mechanism in St. Albans built by Richard of Wallingford, the monumental clock in Strasbourg Cathedral, and the famous astronomical clock in Padua built by Giovanni de'Dondi.[11] The popularity of the mechanical clock spread quickly throughout western Europe. Cities measured their prestige by their elaborate clock towers; references to clocks show up with increasing frequency in the literature of the period.[12]

Clocks were closely associated with miniature mechanical models of the universe that had originated in classical Greece.[13] Using the clock mechanism, these models could be made self-acting so that their motions would match the motions of the actual stars and planets. Many of the early clocks were only incidentally timetelling devices. For example, Dondi's astronomical clock did not simply tell the minutes and hours. On various faces the clock showed the position of the sun, moon, and five other planets.[14] As a model of the universe, the clock played an important role in the development of a mechanical philosophy. Through the clock, people learned to associate the functioning of the universe with the functioning of a machine. As Otto Mayr has concluded, "Clocks, in short, helped to teach Europeans how to think 'mechanically.'"[15]

Renaissance Atomism

The Renaissance saw a revival of interest in the philosophy of the Greek atomists.[16] In the early fifteenth century Italian humanists rediscovered a copy of Lucretius's *De rerum natura* (*On the Nature of Things*) (first century B.C.). According to Lucretius, all natural phenomena, including gases, liquids, solids, and even biological organisms, could be understood in terms of the continual motions of small, unseen, eternal atoms. Although there were difficulties with the poem's explanation of mental activity in relation to fine round atoms, it became very popular during the fifteenth and sixteenth centuries, especially among Renaissance platonists such as Marsilio Ficino.

Atomism was represented by a broad group of theories, some of which held to the belief that atoms were the fundamental indivisible units of matter which moved in a void, while others held that matter could be infinitely divisible but usually existed in the form of particles that could not be easily divided and which moved through a subtle material which filled all space.[17] By the seventeenth century what became known as the corpuscular philosophy began to play a significant role in explaining natural phenomena. For example, Isaac Beekman, who was influenced by Democritus's and Lucretius's theories of atomism and by

Hero of Alexandria's studies of hydraulic and pneumatic machines, argued that the qualities of material bodies arose from the shape, arrangement, and motion of atoms, contrary to the Aristotelian notion that bodies contained a mixture of "essences" (hot, cold, wet, dry). For Beekman, the shapes of atoms determined the qualities of wetness or dryness, while the motion of atoms determined the hotness or coldness of a body.[18] But his work was not widely known during his time.

One of the most important adherents of a corpuscular or atomistic philosophy was Galileo.[19] As a Neoplatonist, Galileo believed that the ultimate reality of the world must be mathematical. According to him the great book of nature "is written in the mathematical language, and the symbols are triangles, circles, and other geometrical figures, without whose help it is impossible to comprehend a single word of it."[20] Galileo's mathematical view of nature led him to distinguish between primary qualities which were objective, mathematical, and unchanging, and secondary qualities which were subjective, sensible, and changeable. Although we experience the secondary qualities of material, such as taste, smell, color, and sound, Galileo argued, these qualities were illusions that could be explained by the primary qualities such as size, shape, quantity, and motion. Through their actions on the senses, the primary qualities could produce the illusions of taste, smell, color, or sound.

The atomistic or corpuscular philosophy provided Galileo with a model for explaining the mathematical character of nature and the distinction between primary and secondary qualities. In his *Dialogues Concerning Two New Sciences* (1638), Galileo combined Democritus's atomism with Hero of Alexandria's pneumatics to explain material bodies.[21] After demonstrating that the force of a vacuum could cause two polished plates of glass to resist being pulled apart, Galileo argues that a similar cause may "explain the coherence of smaller parts and indeed of the very smallest particles of these materials."[22] That is, material bodies are held together by the tiny internal vacua that exist between the particles. Using this model, Galileo was able to explain how heat could cause a solid body to melt by reducing the vacua between the particles,

> The explanation might lie in the fact that the extremely fine particles of fire, penetrating the slender pores of the metal . . . would fill the small intervening vacua and would set free these small particles from the attraction which these same vacua exert upon them and which prevents their separation.[23]

But the size and shape of particles of matter was not enough to explain physical phenomena.[24] According to Galileo, the motion of particles could also have an effect. For example, in explaining how sunlight was able to melt solids, he argued that "combustions and resolutions are accompanied by motion, and that, the most rapid; note the action of lightning and of powder as used in mines and petards."[25] Through the use of an atomistic or corpuscular theory, Galileo was able to propose a world view in which subjective secondary qualities were reduced to the mechanical actions of particles. He concluded,

But that external bodies, to excite in us these tastes, these odours, and these sounds demanded other than size, figure, number, and slow or rapid motion, I do not believe; and I judge that if the ears, the tongue, and the nostrils were taken away, the figure, the numbers, and the motions would indeed remain, but not the odours nor the tastes nor the sounds.[26]

Descartes' Mechanical Philosophy

The atomistic or corpuscular theories of such scientists as Beekman and Galileo were an important step toward a mechanical world view but it remained for these new ideas of nature to be incorporated into a new philosophy. During the middle of the seventeenth century many earlier ideas were synthesized to form what has been called mechanical philosophy. The major figure in the movement was the French philosopher René Descartes. Through his writings, Descartes brought together elements drawn from Platonic philosophy, Greek atomism, and mechanical inventions, and created a new view of the world and the individual's relation to it. His works provided a philosophical basis for the Scientific Revolution and they still play a large role in shaping modern thought about the world.

Descartes was reacting against the philosophy that had dominated Western thought throughout the Middle Ages. Scholasticism had been closely associated with the Church and was primarily concerned with elaborate textual analysis and commentaries on the works of Aristotle. It was a major effort both to reconcile various translations that had come into Western Europe and to reconcile Aristotle, a pagan, with the Christian Church. By the fifteenth century, Scholasticism had become weighted down by its own dogmas. The world was changing. The Copernican revolution in astronomy had shaken the hierarchical system on which much of Scholasticism had been based. The Renaissance's new emphasis on humanism meant a shift away from the sacred and toward the secular. Descartes had been unhappy with the Scholastic method and he had little faith in Aristotelian logic. He wanted to establish a means for acquiring knowledge about the world that was based on truth rather than the unchallenged authority of the Church.

In 1637, Descartes published his "Discourse on the Method of Rightly Conducting the Reason and Seeking for Truth in the Sciences." In this work he attacked the current philosophical systems and described how he found himself doubting whether he could find some principle of certitude on which to base his philosophy. Descartes wanted philosophy to have the same certainty as mathematics.[27] To this end he set down four "precepts of logic." The first was to accept only "what was presented to my mind so clearly and distinctly that I could have no occasion to doubt it."[28] The second was to analyze each problem by dividing it into its parts. Then, thirdly, one could begin with the simplest case and move to more complex ones. And, finally, the solution would be applied to the broadest possible context. These four precepts define the approach to the world taken by mechanical philosophy. Like Platonic philosophy, it was essentially rationalistic—that is, it defined reality in terms of ideas within the mind. And like the atomistic philosophy, it was reductionistic—that is, it reduced complex phenomena to some simple set of concepts.

Using his four precepts of logic, Descartes began his search for a principle of certitude on which to build his philosophy. First, he had to reject everything that there was the least possibility he could doubt.[29] In doing so, Descartes found that there was something that could be doubted about every thought that came into his mind. But in all of this skepticism there was one fact that he could not doubt—the fact that he himself was doing the doubting. Therefore the first principle of Descartes' philosophy became "I think, therefore I am (*cogito ergo sum*)." From this principle he was able to establish an important distinction. Although it was possible for him to doubt the fact that he had a body, he still had to believe that he existed since he was thinking. But if he stopped thinking there was no way to prove to himself that he existed even though his body might still exist. This led Descartes to draw a rigid distinction between the mind, or soul, and the body.[30] According to this dualism, the physical world, everything external to the human mind, is absolutely separate and distinct from the soul or spirit. This position entailed a significant new view of the material world. The matter that composed the physical universe must be devoid of any characteristics associated with soul or spirit. Gravity could no longer be explained in terms of an object's "desire" to return to its natural place. The actions of a magnet could no longer be attributed to the existence of a "magnetic soul."[31] All matter, both organic and inorganic was entirely passive. Only through God's action were spirit and matter brought together in the human mind, but the rest of organic nature, including the human body, was inert.

In 1644 Descartes explained his mechanical philosophy in a work entitled "The Principles of Philosophy." The work was based on the logic that he had developed in the "Discourse." Since the rational mind was the basis of certitude on which he built his philosophy, the test of what was true could be performed by the mind. Whatever ideas were "clear and distinct" must be true. Therefore the goal of Descartes' philosophy was to determine which ideas about the world were clear and distinct, and to explain the processes of the world in terms of those ideas.

What were these clear and distinct ideas? According to Descartes, mathematics, especially geometry, provided the best model of these types of ideas. Like Galileo, he believed that the world of nature could be reduced to concepts describable in terms of mathematics. For Descartes, one such concept was matter. He argued that "the nature of matter or of body in its universal aspect, does not consist in its being hard, or heavy, or coloured, or one that affects our senses in some other way, but solely in the fact that it is a substance extended in length, breadth, and depth."[32] That is, the essential characteristic of matter was extension, which could be described in terms of geometry. Since space was extended, it could not be conceived as distinct from matter. For Descartes, the entire universe was filled with a material, or *plenum,* which only appears to be empty because it is composed of a very subtle material (*matière subtile*). In a literal sense, Descartes was not an atomist, since he held that matter was indefinitely divisible, but he believed that matter usually existed in the form of corpuscles which could not be divided easily.[33]

If extended matter was one pillar of Descartes' world view, the other pillar was motion, which, like matter, was describable in terms of geometry. Since matter was totally devoid of any active qualities such as soul or spirit, its source of motion could not be internal but had to come from outside. Descartes argued that the source of motion in the universe was God, who set matter in motion at the creation.[34] Once matter was set in motion, its motion could not be destroyed; it could only be transferred by impact from one piece of matter to another. (Descartes ignored the problem of explaining how the transfer of motion would be possible in a totally filled universe.) Motion was no longer a special quality that needed to be explained. By postulating a principle of inertia, Descartes could argue that matter will continue in a state of rectilinear motion until some external factor changes it. The law of inertia captured the essence of the new world view. As historian Richard Westfall puts it, "The world is a machine, composed of inert bodies, moved by physical necessity, indifferent to the existence of thinking beings."[35]

Descartes used the mechanical concepts of matter and motion to explain the functionings of the universe. Since the universe was completely filled with a subtle material, he believed that a gross body could move only into a space that had been vacated by some subtle matter and that this subtle matter, through its motion, will eventually cause some other subtle matter to move into the space vacated by the original body. As a consequence of the motions of gross bodies, the subtle matter will move in a complete circuit and start to form a vortex or whirlpool motion. Since a large body would be carried in a complete circuit by the vortex motion, Descartes believed that such a motion could explain the motions of the planets around the sun. He also believed that a smaller vortex around the earth could account for the action of gravity in the same way that a whirlpool of water can pull floating objects toward its center.

In place of occult powers and desires, Descartes substituted the mechanical actions of inert matter. Rather than attributing magnetic phenomena to the existence of an animate soul, he attempted to explain magnetic attraction and repulsion in terms of matter and motion. According to Descartes, through the action of the vortex, a magnet emitted screw-shaped particles which fit exactly into screw-shaped pores in iron.[36] As the particles were twisted into the pores, the iron would be drawn to the magnet like a nut on a bolt. By postulating the existence of right-handed and left-handed particles, he was able to explain the fact that a magnet had a north and a south pole, each of which would attract an unlike pole and repel a like pole.

Although many of Descartes' explanations of physical phenomena, especially his reliance on the vortex, were rejected by later scientists, his philosophy became the cornerstone of the mechanical world view. He replaced the world of the senses with an ideal world based on geometric quantities and provided a philosophical foundation for Galileo's distinction between primary and secondary qualities. Descartes' dualism between mind and body separated the primary qualities of extension and motion from the secondary qualities of sensation, which "can be representative of nothing that exists out of our mind."[37] More and more the world was seen as being removed from our actual experiences by functioning almost like a giant machine.

Gassendi's Mechanical Philosophy

A second major figure in the creation of a mechanical philosophy was the French priest and scientist Pierre Gassendi.[38] During the middle of the seventeenth century he turned to the atomistic philosophy of Epicurus and Lucretius as a way of developing a new non-Aristotelian natural philosophy. Gassendi's atomism became the most significant alternative to Descartes' philosophy. The major difference between the two philosophies was that Gassendi did not believe that matter filled all of space or that, like space, matter was infinitely divisible. As an atomist, he held that matter existed in the form of small indivisible units. Although Gassendi did not equate space with matter, he shared with Descartes the belief that all physical phenomena could be explained in terms of matter and motion. Once the material atoms had been created and set in motion, all natural phenomena were the result of the collisions between atoms moving in a void. The properties of bodies depended on the interactions and motions of atoms which came together to form corpuscles of varying shapes and sizes.[39] If the corpuscles were smooth, the body would be a fluid. Gassendi believed that both heat and cold were associated with specific types of atoms. A body would be hot if it contained "calorific atoms" which were small, round, and easy to move, while a body would be cold if it contained "frigorific atoms," which were heavy and difficult to move.[40]

Although Gassendi's mechanical philosophy was influenced by Epicurean atomism, he rejected the atheistic implications that had made atomism unacceptable to many European philosophers. Unlike Epicurus, Gassendi argued that the atoms had not eternally existed but were created by God. Like Descartes, he argued that it was God who gave matter its initial motion. Also, unlike Epicurus, but similar to Descartes, Gassendi believed that the human soul was immaterial and could not be explained in terms of material atoms.[41]

Hobbes's Mechanical Philosophy

The mechanical philosophies of both Descartes and Gassendi had a great impact on late seventeenth-century thought. Their ideas were particularly influential among a group of scientists and philosophers centered around Sir Charles Cavendish and his older brother William Cavendish, Marquis (later Duke) of Newcastle.[42] The Newcastle Circle was loyal to the Royalist cause, and during the English Civil War the group spent several years exiled in Paris where they became closely associated with Marin Mersenne, Descartes' secretary, and with Gassendi. The Newcastle Circle returned home after the death of Charles I and played an important role in the transmission of mechanical philosophy into England.

The intellectual center of the Newcastle Circle was Thomas Hobbes, who had contacts with Galileo, Descartes, and Gassendi. Although better known today for his political philosophy, especially the *Leviathan,* Hobbes was a major figure in the establishment of mechanical philosophy.[43] During the 1640s and 1650s, he formulated a mechanical explanation of natural phenomena. He was one of the first mechanical philosophers to emphasize the role of the motion of matter rather than its size and shape.[44] For Hobbes the motion of any body was the result of the actions of some other body that was already in motion, including a body at rest. He said, "The cause of all motion and change is motion."[45]

In 1655 Hobbes published *De Corpore,* in which he abandoned his earlier belief that particles moved in a void and instead argued for the existence of a fluidlike material aether that pervaded all space between the particles and provided a mechanical connection between them.[46] Although matter played an important role in all physical phenomena, Hobbes continued to emphasize the role of motion. For example, he argued that odors were caused by moving particles within bodies transmitting their motion through the aether to the nose, while most other mechanical philosophers explained odors in terms of the shapes of particles given off by a body.

One of Hobbes's most important contributions to mechanical philosophy was his theory of hardness and cohesion. At various times he postulated at least three basic reasons to account for the hardness of bodies. Like Galileo, he argued that bodies would be hard if the constituent particles cleaved together so that pulling them apart would be difficult.[47] Hobbes also argued that the hardness of bodies was the result of the packing together of particles, which he believed could have different degrees of hardness.[48] But in his later works Hobbes explained hardness in terms of the motion of particles; in doing so, he put forward one of the first truly kinetic theories of matter in the seventeenth century.[49] He argued, "For the cause therefore of hardness, I suppose the reciprocation of motion in those things which are hard, to be very swift, and in very small circles."[50] Since Hobbes, like Galileo, believed that circular motion was natural, the hardness of a body depended on the velocity of the particles and the size of the circles.[51] The hardest bodies were those in which the motion was the fastest and the circles were the smallest.[52] Heat would soften a body because the motion of the particles of fire would cause the particles of the body to move in larger circles and thereby reduce its hardness. He argues that a bow that is bent would return to its original shape because "the particles of the bended body, whilst it is held bent, do nevertheless retain their motion; and by this motion they restore it soon as the force is removed by which it was bent."[53] Hobbes applied this kinetic model to a wide range of problems in the area of elasticity and the strength of materials.[54]

Hobbes revived the theological attacks on atomism by calling attention to its atheistic implications. In attempting to eliminate all occult qualities from nature, Hobbes denied the existence of any immaterial substances.[55] He said, "By the name *spirit,* we understand a *body natural,* but of such subtlety that it worketh not upon the senses; but that filleth up the places which the image of a visible body might fill up."[56] Hobbes rejected the Cartesian dualism between mind and matter and instead argued that "mind will be nothing but the motions in certain parts of an organic body."[57] Since the Bible states that the soul exists within the human being, Hobbes argued that such "words do imply *locality;* and locality is *dimension;* and whatsoever hath dimension is *body* be it ever so subtle."[58] Hobbes was attacked as a heretic, and many other supporters of mechanical philosophy sought to disassociate it from any atheistic interpretations.

Boyle's Mechanical Philosophy

Descartes, Gassendi, and Hobbes were the great systematizers of mechanical philosophy but their theories were hypothetical and none of them provided much experimental evidence. One of the leading scientists who used experiments to defend mechanical philosophy was the chemist Robert Boyle, an early member of the Royal Society of London.[59] Through his association with members of the Newcastle Circle, Boyle came to be influenced by the theories of Descartes and Gassendi. In fact, he seemed to make little or no distinction between the plenist theory of Descartes and the atomistic theory of Gassendi. Boyle incorporated both Cartesian and atomistic explanations into what he called his "corpuscular philosophy" and argued that all phenomena could be explained in terms of matter and motion.[60] Although primary matter was infinitely divisible by God, it existed in the form of extended, impenetrable particles or corpuscles. The secondary qualities of taste, color, and odor could be explained in terms of the size, shape, arrangement, and motion of the corpuscles of primary matter.

Boyle's most important contribution to the establishment of a mechanical philosophy was experimental, not theoretical.[61] He was not concerned with proving that a particular theory, such as Descartes' or Gassendi's, was correct; rather he wanted to provide experimental evidence that natural phenomena could be explained by *some* mechanical hypothesis. He said,

> That then, which I chiefly aim at, is to make it probable to you by experiments (which I think hath not yet been done), that almost all sorts of qualities, most of which have been by the schools either left unexplicated, or generally referred to I know not what incomprehensible substantial forms, may be produced mechanically; I mean by such corporeal agents, as do not appear either to work otherwise than by virtue of the motion, size, figure, and contrivance of their own parts (which attributes I call the mechanical affections of matter, because to them men willingly refer various operations of mechanical engines).[62]

With the scientific instruments available in the seventeenth century, Boyle was not able to provide experimental proof that atoms, or corpuscles, existed, but he did help to convince many of his contemporaries that the results of experimental research were consistent with a mechanical explanation of nature. In many experiments his main purpose was to cast doubt on the Aristotelian notion that qualities, such as hotness, coldness, sweetness, sourness, redness, or blueness, were essential properties of matter. Many of Boyle's experiments were described in his *Sceptical Chymist* (1661) and *Experiments, Notes, etc. about the Mechanical Origins or Production of Divers Particular Qualities* (1675).[63]

In one series of experiments Boyle demonstrated that taste could not be an inherent quality associated with particular materials. As evidence he was able to produce a very spicy tasting substance by mixing together two very bland substances. He was also able to make a bland substance (saltpeter) taste fiery by a process of distillation. Boyle did not believe that Aristotelians could explain the emergence of a quality from a material that did not already possess that quality, but such results could be explained mechanically. When saltpeter was distilled by fire, he argued,

> We should suppose the corpuscles of nitre to be little prisms, whose angles and ends are too obtuse or blunt to make vigorous and deep impressions on the tongue; and yet, if these little prisms are by a violent heat split . . . they may come to have parts so much smaller than before, . . . the sharpness of their sides and parts may fit them to stab and cut [the pores of the tongue].[64]

In other experiments he combined chemicals at room temperature and produced a mixture that was either hotter or colder than the original substances. He also demonstrated that the combination of an acidic substance and an alkalisate salt produced saltpeter, whose properties differed from each of the combined substances.[65] All of these experiments seemed to undermine the Aristotelian notion of qualities, and helped to strengthen the scientific appeal of mechanical philosophy.

Boyle not only tried to provide an experimental basis for mechanical philosophy, but to place it within an acceptable theological framework.[66] For Boyle, the orderly behavior of the mechanical world did not imply an atheistic materialism. Based on his concept of natural law, he argued that only an intelligent agent could obey a law, but the material corpuscles, or atoms, that composed the universe were inanimate and could not, by themselves, obey the mechanical laws which governed the world. In Boyle's view mechanical philosophy necessitated the existence of some designer who created the order of the world. Instead of implying atheism, he argued that mechanical principles were the "alphabet in which God wrote the world."[67] Although God could intervene in the world through miracles to correct any irregularities, Boyle believed that most of the time there was no need for such extraordinary intervention. In general, he argued, the universe was

> like a rare clock, such as may be that at Strasburg, where all things are so skillfully contrived, that the engine being once set a-moving, all things proceed according to the artificer's first design, and the motions . . . do not require the peculiar interposing of the artificer . . . but perform their functions upon particular occasions, by virtue of the general and primitive contrivance of the whole engine.[68]

By transforming God into a clockmaker or engineer, Boyle provided a new theological framework for mechanical philosophy.

Boyle's experiments and his theology helped to promote the acceptance of mechanical philosophy, but there were still some problems that had to be faced. Several types of physical phenomena were very difficult to explain using only matter and motion.[69] One of these was gravity. The action of gravity on a body caused it to undergo uniform accelerated motion. How could the collisions of particles moving with a constant velocity cause a body to move with uniform acceleration? Descartes had attempted to solve the problem of gravity by postulating the existence of matter in vortex motion around the earth, which pushed bodies swimming in the whirlpool toward the center. Although Descartes' theory provided a qualitative explanation of gravity, it failed in explaining the details of gravity. For example, the theory predicted that bodies would fall toward the earth's axis rather than the earth's center so that there would be no gravity at the north and south poles. Other phenomena, such as the cohesion of bodies, capillary action, electrical and magnetic attraction and repulsion, the expansion of gases, and chemical affinities, were also difficult to explain, even qualitatively, in terms of matter and motion.

During the second half of the seventeenth century another, more serious weakness in mechanical philosophy was becoming apparent to many scientists. Although mechanical philosophy was based on mathematical or geometric concepts and principles, most mechanical theories gave qualitative rather than quantitative results. Acids might be explained as sharp wedge-shaped particles, but no seventeenth-century chemist was able to discover the precise angle of a corpuscle of sulphuric acid or to predict the chemical reaction that would take place if the corpuscle had a thirty-degree angle instead of a twenty-degree angle. When a few scientists did attempt to apply the mathematical laws of mechanics to physical phenomena, they found, in many cases, that the results contradicted mechanical philosophy. For example, Descartes and others had based much of their mechanical philosophy on the premise that the ''quantity of motion'' (mass times velocity) was conserved during the impact between two pieces of matter, but the Dutch scientist Christiaan Huygens discovered that the ''quantity of motion'' was not always conserved in every type of collision.[70]

The Newtonian World View

By the beginning of the eighteenth century, the mechanical world view had reached a new stage of development with the work of Sir Isaac Newton, who solved most of the remaining problems that had faced mechanical philosophy.[71] As a student he read the works of Descartes, Gassendi, Hobbes, and Boyle, and in one of his first published papers, ''An Hypothesis Explaining the Properties of Light'' (1672), Newton adopted a corpuscular theory, explaining optical phenomena in terms of the motion of corpuscles of light passing through a fluidlike aether composed of smaller particles. But he found it difficult, if not impossible, to develop a quantitative, universal theory of natural phenomena, such as gravity, using traditional mechanical philosophy. In some of his early papers Newton tried to analyze a body moving in a circular orbit by postulating a series of impacts which deflected the body into a circular path. But such a mechanical model provided him with only an approximate correlation between the orbit of the moon and the accelerated motion of falling bodies on the earth. Newton realized that a universal theory of gravitation required an exact mathematical correlation between the motion of celestial objects, like the moon, and the motion of terrestrial bodies, like a falling apple. To accomplish this task, Newton would have to modify accepted mechanical philosophy.

Newton proposed his revolutionary reformulation of mechanical philosophy in *Philosophae naturalis principia mathematica* (1687). Throughout the *Principia* he continued to assume that the universe was composed of corpuscles of matter that were hard, extended, impenetrable, and endowed with inertial motion.[72] But Newton's corpuscles did not interact with each other through collisions or impact. Instead, he argued that a corpuscle of matter contained "a certain power diffused from the centre to all places around to move the bodies that are in them."[73] This power revealed a new definition of force. Previously mechanical philosophers had assumed that force was simply the result of a moving piece of matter impinging on another, what Descartes called the "force of a body's motion." But Newton introduced a new concept of force, one which gave it an independent standing in mechanical philosophy along with matter and motion. According to him the world was not composed of just matter and motion, but matter, motion, and attractive and repulsive forces that acted between bodies. These forces, which he was able to describe mathematically, had the peculiar property of acting on another material body across an empty space. Throughout the *Principia* Newton refused to speculate on the cause of action-at-a-distance and seemed willing only to define forces in terms of a mathematical description of a tendency of two bodies to move together or apart. But he was able to demonstrate that his mathematical concept of force could be applied, with great success, to the problems that had been facing mechanical philosophy.

In Book One of the *Principia,* Newton applied his concept of force to a simple system composed of an ideal point mass in orbit around a center of attracting force. In such a system he was able to demonstrate that Kepler's three laws of planetary motion would hold if the attractive force varied as the inverse square of the distance between the point mass and the center. In Book Two, Newton extended his analysis from ideal point masses to bodies moving through a resisting medium such as a fluid. Through a detailed series of mathematical arguments, he was able to show that a body in a medium moving in a vortex, such as postulated by Descartes' mechanical philosophy, would not follow Kepler's laws. Even more devastating to the Cartesian system, Newton was able to demonstrate that a vortex would dissipate unless it was maintained by some external force. In Book Three, Newton applied his concept of force to the "system of the world." By postulating that a universal attractive gravitational force acted between all particles of matter in the universe, he was able to demonstrate that the forces which kept the planets in their orbits around the sun and which kept the moon in its orbit around the earth, were mathematically identical to the force that caused an apple to fall to the ground on the earth.

In the *Opticks* (1704), Newton extended his concept of force to microscopic phenomena. In what became known as Query 31, Newton argued,

Have not the small Particles of Bodies certain Powers, Virtues, or Forces, by which they act at a distance . . . upon one another for producing a great Part of the Phaenomena of Nature? For it's well known, that Bodies act one upon another by the Attractions of Gravity, Magnetism, and Electricity; and these Instances show that there may be more attractive Powers than these.[74]

Throughout the long query, he gives examples of phenomena that had been difficult to explain in terms of matter and motion but which could be understood in terms of various attractive and repulsive forces acting between particles. For example, Newton described several types of chemical reactions such as the absorption of water from the air by salt of tartar, which then will not release the water unless a great deal of heat is applied. He asks, "Is not this done by an Attraction between the Particles of Salt of Tartar, and the Particles of the Water which float in the Air in the form of Vapours?"[75] The release of heat when acid came in contact with iron filings was explained by an attractive force causing the particles of acid and the particles of iron to move toward each other at great speeds.[76] Other problems, such as the expansion of gases, evaporation, and fermentation, could be explained by the existence of repulsive forces acting between particles.[77] Although Newton was unable to formulate microscopic laws as precise or universal as his law of gravitation, his definition of force as a change in the quantity of motion provided a mathematical framework for the discovery of such laws.

With Newton, mechanical philosophy was transformed to include the concept of force along with matter and motion. But his notion of force led to a debate over the philosophical basis of mechanical philosophy. The idea of action-at-a-distance led some critics to accuse Newton of reintroducing occult qualities into nature. In his public writings Newton never discussed the exact nature of force, willing only to describe it in terms of a mathematical definition. He attempted to reduce gravitational force to a more traditional mechanical explanation by postulating the existence of an all pervasive aether, but this simply moved the problem to another level since the all pervasive aether was composed of repulsive particles. For Newton the ultimate cause of force was theological and not mechanical.[78] The existence of force provided him with the evidence that God was continually active in the universe. Newton associated infinite space with God, calling the universe the sensorium of God.[79] According to this view, God perceives material objects the same way the human mind perceives ideas.[80] In humans, thought and actions are distinct, whereas God has only a mind, or sensorium, so that will and action are one and the same. For Newton, such a theological position resolved the philosophical problem of action-at-a-distance. Since all bodies exist in the sensorium of God, it is possible for God to act on all bodies at the same time. In this view force becomes a manifestation of the will of God acting in the universe.

Newton's theological interpretation of force led to a philosophical debate concerning the role of God in the mechanical world view. Newton believed that God's continual action would be required to keep the world running smoothly. He mistakenly thought that the highly eccentric orbits of the comets would perturb the orbits of the planets "which will be apt to increase, till this System wants a Reformation."[81] Therefore, the world required "a powerful ever-living Agent, who being in all Places, is more able by his Will to move the Bodies within his boundless uniform Sensorium, and thereby to form and reform the Parts of the Universe."[82] To some it seemed as if mechanical philosophy had taken on some vitalistic overtones. Newton's position was challenged by several scientists and philosophers, the most vocal of whom was Leibniz. He was particularly opposed to the idea that Newton's mechanical world view required God to intervene continually in the world. Leibniz said,

Sir *Isaac Newton,* and his Followers, have also a very odd Opinion concerning the Work of God. According to their Doctrine, God Almighty wants to *wind up* his Watch from Time to Time: Otherwise it would cease to move. He had not, it seems, sufficient Foresight to make it a perpetual Motion. Nay, the Machine of God's making, is so imperfect, according to these Gentlemen, that He is obliged to *clean* it now and then by an extraordinary Concourse, and even to *mend* it, as a Clockmaker mends his Work.[83]

Leibniz believed that an all-powerful God should be able to create a clockwork universe that was perfect. Newton's theory seemed to limit God's creative power to create a self-maintaining clockwork.

Newton was reticent to enter into philosophical debates concerning his theory but his pupil and friend, Dr. Samuel Clarke, defended Newton's position in a long-written debate with Leibniz.[84] According to Clarke, the Newtonian view did not belittle God; rather it indicated that God, through His continual presence, had an active interest in the world. In fact, Clarke argued that Leibniz's concept of God would lead to materialism and atheism since God would only be the prime mover who begins the universe and then withdraws outside the world. But, for Leibniz, God was not absent from the world; as the supremely rational Being, God was constrained to act through the mechanical laws that He created for the universe. Once the mechanism of the universe had been created, God could not tamper with or break His own laws.

The Leibniz-Clarke debate continued without any real resolution and ended only when Leibniz died. But in the years following the debate, each side won some arguments and lost others. More and more, the Newtonian system of natural philosophy, with its concept of attractive and repulsive forces, overcame its critics and became the method by which scientists explained the world. Although the Newtonian system was victorious in the course of the battle, the character of the concept of force underwent change. For Newton, the introduction of force was proof that a purely material system of matter and motion was insufficient for explaining the world. But even as the system became accepted, most scientists rejected Newton's theological interpretation. Although natural philosophers could not yet explain force in terms of matter and motion, many assumed that a mechanical explanation would be found in the future. In the meantime, they could solve problems by treating them *as if* an attractive or repulsive force acted between bodies. Other natural philosophers came to view force simply as another natural property of matter along with extension, hardness, impenetrability, and inertia.

The Clockwork Universe

Ironically the success of Newtonian natural philosophy led to the discovery of evidence to support the Leibnizian clockwork.[85] By the eighteenth century, natural philosophers were discovering that the clockwork universe was much more self-regulating than Newton had believed. Newton had noticed that the planets sometimes strayed from their predicted orbits. This led him to believe that God had to intervene and correct the planet's path—resetting the clock. Using an improved version of the calculus, several French mathematicians and scientists, including Jean le Rond d'Alembert, Alexis Claude Clairault, Joseph Lagrange, and Pierre Simon de Laplace successfully applied Newtonian mechanics to many of the "irregularities" that had concerned Newton.[86] In particular, Laplace was able to demonstrate that the Newtonian universe was self-correcting. He was able to prove that if the solar system was governed by Newton's force of gravitation, any planet that strayed out of its orbit, because of some perturbation, would experience either a greater or weaker gravitational force that would return it back to its correct orbit. When Napoleon asked about the place of God in his work, Laplace replied, "Sire I have no need for that hypothesis."[87] Laplace and other scientists came to see Newton's inverse square law of gravitation as the model for all other forces. The laws of electricity and magnetism, along with the intensity of light and radiant heat, all seemed to follow an inverse square law.[88] This mathematical character of the laws of force led Laplace to conceive of a completely deterministic world. In 1812 he speculated that if an omniscient calculator could know the positions and the velocities of all of the particles in the world at a given instant, this calculator would be able to know all past and future events with an absolute certainty. By the beginning of the nineteenth-century, Newtonian force had become part of mechanical philosophy but so had the Leibnizian self-regulating clockwork.

Mechanical Theories of Chemistry and Heat

In the nineteenth century, mechanical philosophy came to explain more and more about the nature of matter. By the first part of the century, chemists were beginning to extend Newton's idea of attractive and repulsive forces into the notion of chemical atoms.[89] In 1808 John Dalton discovered that chemicals combined together in definite proportions and that these proportions were multiples of each other.[90] Based on this discovery, he theorized that the individual elements that chemists had identified were atoms and that chemical compounds were the combination of several types of individual atoms. Using fixed proportions and the idea of chemical atoms, Dalton was able to make some estimates of the weight of the different atoms relative to each other (although Dalton's belief that water was composed of one hydrogen and one oxygen atom introduced an error into his table of weights).

One of the greatest successes of nineteenth-century scientists was the extension of mechanical philosophy to the theory of heat.[91] It was the mechanical theory of heat that secured atomism as a fundamental element of science. Most eighteenth-century scientists had believed that heat was a material substance—an imponderable fluid that was called caloric. The transfer of heat from one body to another was thought to involve the transfer of a certain amount of fluidlike caloric. But, by 1800, some scientists began to question the concept of caloric. The American expatriate Count Rumford (Benjamin Thompson) showed that simple friction could generate an unlimited quantity of heat, raising the question of the source of such great amounts of caloric when the actual material bodies producing the friction were fixed in size. Also, in 1848 the English scientist J. P. Joule used a set of paddles rotating in a tub of water to show that mechanical work could be converted into heat at a fixed rate.

By the second half of the century, several scientists, such as James Clerk Maxwell, Rudolf Clausius and Ludwig Boltzmann, argued that heat was not a material substance, but rather the mechanical motions of the atoms and molecules of matter. According to this theory, heat transfer was simply the transfer of atomic and molecular motion, through collisions, from one body to another. The mechanical theory of heat helped to explain several physical problems such as the diffusion of gases and the theory of thermodynamics. Also, since the weight, size, and number of physical atoms in the mechanical theory of heat correlated almost exactly with the properties of Dalton's chemical atoms, this theory helped to unify physical and chemical atomism and brought to fulfillment the Newtonian view of matter.

Conclusion

Throughout its history the mechanical world view underwent development and change. By the nineteenth century several problems had begun to modify some of the more extreme versions of mechanical philosophy and bring it closer to the opposing organic world view. The continual problem of explaining the nature of Newtonian forces acting-at-a-distance was difficult, if not impossible, to reduce to the effects of matter and motion. Even Newton's own attempt to reduce gravitational attraction to the action of a space-filling material aether required repulsive forces to explain the aether. Forces acting between particles seemed to reintroduce "occult qualities" into nature by giving matter some active principle. In the middle of the eighteenth century, some Newtonians began to reduce physical phenomena to the presence or absence of subtle or imponderable fluids.[92] In this way, they were able to evade some of the more troubling problems of reductionism, by resorting to a form of quantified materialism. This was especially true in the area of electricity and magnetism which always had been difficult to explain in terms of matter and motion. For example, Benjamin Franklin believed that all bodies contained a natural amount of electrical fluid and that the addition or subtraction of this electrical fluid would leave a body with a positive or negative charge.[93]

Other Newtonians attempted to solve the problem of mechanical philosophy by reducing all physical phenomena to the actions of forces. For example, Roger Joseph Boscovich believed that nature could be conceived as a system of geometric points surrounded by one continuous set of attractive and repulsive forces.[94] None of these versions of mechanical philosophy could be called organic, but each helped to make the opposition between the two world views less extreme.

In spite of this modification and change, one set of ideas serves to characterize the mechanical world view in all of its various forms.[95] In general, this view assumes that the world functions like a machine, particularly a clock. Phenomena can be explained in terms of homogeneous, inert, passive pieces of material that can interact with other pieces of matter through contact or through forces, so that motion can be transmitted from one piece to another. In this world, the only fundamental change is that which results from the different arrangements and different motions of pieces of matter. Since it is assumed that pieces of matter interact with each other according to predictable, mathematical laws, all phenomena could be understood through a process of reductionism in which complex problems are solved by breaking them into smaller and smaller parts and then analyzing those parts. In such a world view a total understanding of the behavior of the whole can be gained through an understanding of each of its parts.

The mechanical world view was applied not only to physical phenomena but to organic phenomena as well. Within this world view there was no fundamental conflict between technology and the organic. Along with the motion of the planets, the functioning of machines, and the action of matter, mechanical concepts could also explain respiration, circulation, digestion, reproduction, sensation, and even thought itself. Within the mechanical world view, the symbol of the machine transformed life into what could be called mechanical organisms.

3

Mechanical Organisms:
From Automata to
Clockwork Humans

> For what is the *heart*, but a *spring;* and the *nerves*, but so many
> *strings;* and the joints, but so many wheels, giving motion to the
> whole body, such as was intended by the artificer?
>
> THOMAS HOBBES, *Leviathan*

The mechanical world view provided a model for understanding organic life that differed from the model provided by the organic world view. According to the mechanical view, biological phenomena could be explained by reducing them to the actions of the same inert passive material that composed the rest of the world. The machine, with a precise and rational relationship among its parts, became the model for this mechancial world. At first the machine served only as an analogue for biological processes; the organic world could be understood by comparison with some well-known mechanical device or technological process.[1] But as mechanical philosophy became successful as a method of explanation, people no longer saw the machine as simply an analogue for life—life became literally mechanical. They believed that biological processes, such as digestion, respiration, movement, and sensation, were, in fact, technological processes. Although the body might not appear as a machine to our senses, people believed that it must function according to rational, mechanical laws.

Automata

Mechanical theories of life became popular during the seventeenth-century Scientific Revolution, but, similar to mechanical philosophy, earlier ideas influenced the development of a mechanical concept of life. Of unmistakable importance, for example, was the considerable interest, throughout history, in automata.[2] Humans have always had a deep urge to simulate living things by means of art or technology. We have evidence of articulated figurines from Egyptian times that could be moved by strings or by hand, while some statues had concealed speaking tubes enabling them to appear to speak. The exact function of the figures is not clear, but some religious significance is plausible, especially in rituals where life would be recreated. A belief has persisted since ancient times that the magical induction of spirit could bring life to an inanimate statue or figurine representing a god. There is also evidence, during the Greco-Roman period, of the presence of articulated statues powered by some technological means. Aristotle must have been familiar with automata since he claimed, in *On the Motion of Animals,* that the movements of animals were similar to those of automatic puppets. It was rumored that Mark Anthony used an automaton during his funeral oration for Julius Caesar; the automaton, resembling Caesar, arose from the bier, inciting the crowd to riot.[3]

Greek automata were epitomized in the work of Hero of Alexandria (c. 62 A.D.), who described the ideas of other inventors in his treatise the *Pneumatics.*[4] Certainly on one level, Hero's inventions display the enduring interest in simulating life technologically. For example, through the use of a simple siphon hidden inside a statue of a horse, the animal would appear to drink from a pan of water placed before it. Or, water moving through pipes could be used to force air past a whistle, making birds appear to sing, and the weight of the water could be used to make figurines turn or move around a base.[5] It is not clear whether any or all of these fanciful inventions were actually made, but Hero's intricate drawings show that he understood how technology could be used to simulate life.

Clearly the *Pneumatics* had a great influence on later inventors. Islamic water clocks featured mechanical birds which announced the hours, and by the Middle Ages moving mechanical statues graced many of the new cathedral clocks.[6] The natural association of automata and clocks has been drawn by historian Derek Price;[7] both were based on similar technological developments, first water power and, later, mechanical power. But, more importantly, each represented an urge to create a universe, on the one hand the motion of the sun, moon, and planets, and on the other hand the motion of animate life.

In the later part of the Middle Ages and early Renaissance, the development of the mechanical clock in Europe led to a resurgence in automata. At first clocks simply reminded a human bell ringer when the hours had to be struck, but very soon mechanisms were invented to strike the hours automatically. In many of the clocks, the hours were struck by mechanical figures or "jacks."[8] As other parts of the hour were struck, new jackworks were added, such as birds chirping or roosters crowing. Eventually entire scenes or small plays were acted out by a series of automata. Historians Alfred Chapius and Edmond Droz described the town clock at Nirot, France built in 1570.[9] The clock had fourteen angels striking a hymn on a carillon while the figure of St. Peter moved out of a door and pointed to another door from which the twelve apostles appeared carrying hammers. Depending on the hour, the appropriate number of apostles would strike the bell and return to their door, which St. Peter then closed. At certain times the sky opened up and the Holy Ghost descended to a figure of the Virgin Mary. Another clock with elaborate jackwork was the great cathedral clock in Strasbourg, which contained several mechanical figurines and a naturalistic rooster which crowed the hours, all linked to an elaborate model of the sun, moon, planets, and stars. This type of clock was quite common throughout Europe during the fifteenth and sixteenth centuries and would have encouraged people to draw a connection between technology and animate life.

Apart from clockworks, automata were a subject of interest in their own right and were used for a wide range of functions. There are legends that the great theologian Albertus Magnus constructed a mechanical man out of metal, wax, and leather. Supposedly, his famous pupil, St. Thomas Aquinas, destroyed one of Alertus's works when its babbling irritated him.[10] St. Thomas, who saw little difference between technology and some forms of life, believed that animals could be considered machines since they exhibited regular motion.[11] The British historian Frances Yates has shown that automata were widely used in Elizabethan theaters to create various effects on stage.[12] At the beginning of the seventeenth century, the French engineer Salmon de Caus described several grottoes and moving statues that he created for the Elector of the Palatine.[13] In these grottoes, mythological figures such as Pan, Venus, or Cyclops moved, turned, and played music, all powered by moving water. Such waterworks were popular throughout France and reinforced the belief that life could be explained by technology.

A New Physiology—
The Circulation of the Blood

At the same time that automata were becoming more sophisticated and more popular, changes were taking place in the study of organic life. During the sixteenth and seventeenth centuries, physiology underwent a revolution. Throughout the Middle Ages, medicine had been dominated by the theories of Galen, who believed that the body contained four humors (blood, phlegm, yellow bile, and black bile) whose proper balance insured health. In addition to these humors, Galen thought that life came from a World Soul which entered the body through the lungs in the form of a vital spirit (or *pneuma*). In the lungs it mixed with the arterial blood and was carried to the rest of the body. On the other hand, nutrients went from the digestive tract to the liver where they were transformed into venous blood and were sent, by the heart, to the rest of the body. According to Galen the blood did not circulate from the veins to the arteries; each type of blood was distinct and had its own distribution system.

By the Renaissance, many of Galen's ideas were being challenged. The Italian physiologist, Andrea Vesalius, discovered through dissection that much of Galen's physiology was wrong, and published his findings in 1543 in the work *On the Fabric of the Human Body*. Like other areas of thought, medicine and physiology became caught up in the emerging mechanical world view. After Vesalius's work, it was clear that the old model of physiology was wrong and, for many scientists and philosophers, automata provided a new model for organic life.

One of the most important medical figures in the new physiology was William Harvey, who (though usually credited as the founder of modern medicine) was in many ways more of a transitional figure, combining older attitudes with new theories.[14] He was born in England and studied medicine at the University of Padua, the great Italian university which had been home to Vesalius and Galileo. On his return to England he practiced medicine at St. Bartholomew's Hospital and later became physician to both James I and Charles I.

Harvey in many ways was still rooted in the physiological tradition of Galen, and many of his observations confirmed the traditional views. It was only when direct experimental observation contradicted the ancient authority that Harvey put forth a new theory. In 1628 he published his famous work *On the Motion of the Heart and Blood*. Although much of the work showed a commitment to the older physiology, it introduced a new idea that eventually led to the replacement of Galen's system. Harvey argued that the blood vessels which joined the heart to the lungs are not independent from the blood vessels which lead to the rest of the body. Rather, both constitute a single system. Blood circulates from the heart to the lungs, where it absorbs air and turns bright red, and then returns to the heart, which pumps it through the arteries to the rest of the body, where it returns by the veins to the heart.

Harvey supported his idea of circulation with two observations. First, like Vesalius, he observed that there were no holes in the central wall of the heart; Galen had argued that blood flowed back and forth through the septum of the heart. But Harvey's most important argument for the circulation of the blood was quantitative. Calculating the quantity of blood pumped by the heart in half an hour, Harvey was able to show that the amount was greater than the total quantity of blood in the body. This discredited Galen's theory that the blood was created in the liver and consumed by the organs. One element in Harvey's theory, however, had to be hypothesized—the link between the arteries and the veins; it was not until the invention of the microscope that the capillaries could be seen.

Harvey's theory of circulation required a new physiological model. Galen believed that the walls of the arteries contained a "pulse making force" which moved the blood. Harvey, on the other hand, argued that the pulse was the result of the mechanical impulse of the blood as it was pumped into the arteries by the heart.[15] In fact, it appears that Harvey thought of the circulatory system as analogous to a hydraulic system of pipes, with the heart acting as a pump. In his Lumleian Lectures of 1616 he said, "From the structure of the heart it is clear that the blood is constantly carried through the lungs into the aorta as by two clacks of a water bellows to rayse water."[16] A clack was a leather valve used in pumps that allowed water to flow in only one direction. (As historian George Basalla has noted, the water bellows referred to by Harvey were common devices, similar to the blacksmith's air bellows, which were used to empty moats or fill wine vats.)[17]

Later, in 1649, Harvey drew another analogy between the heart and a pump. In a letter to Jean Riolan, he said,

When water is forced up to a height through lead pipes, by the force and stroke of a *sipho* (a fire engine pump), we are able to distinguish and observe a sequence of events. . . . It is noted in the case of water that there is a continual outflow, although it sometimes shoots further, sometimes nearer and it is so in arteries.[18]

Charles Webster has argued that the term *sipho* referred to fire engine pumps which were well known in London between 1625 and 1660.[19] Unlike the simpler bellows pump, the *sipho* pumped water continuously and was not empty during part of its cycle.

Although Harvey used mechanical analogies to explain his physiology, they were not the central focus of his work. Pumps and pipes could be used to explain the structure of the circulatory system, but Harvey used them primarily to illustrate that each organ was designed for a specific purpose. The idea of a pump might explain the structure of the heart but the source of the actual contractions had to be explained in terms of inherent vital activity. That is, Harvey might argue that the body functioned *like* a machine but he would never have supported the idea that the body *was* a machine.[20] Even if Harvey did not create a new model of mechanical physiology, his theory set the stage for a new theory of organic life.

Philosophical Physiology

One of the most important figures in establishing a mechanical theory of organisms was not a physiologist but the philosopher René Descartes. As we saw in the last chapter, Descartes was the founder of mechanical philosophy, so it was natural that he would explain organic processes in terms of mechanical concepts. If Harvey provided an experimental basis for mechanical organisms, Descartes provided their philosophical foundation.[21]

Descartes' philosophical dualism, which posited a radical distinction between mind and matter, formed the basis of his physiology. Since matter was completely distinct from anything spiritual, the human body could be treated as entirely independent from the soul. The rational aspects of the mind would depend on the soul, but all of the other normal functions of organisms were bodily processes that could be described by mechanical principles.

In his "Discourse on the Method" (1637), Descartes argued that all bodily processes could be understood in terms of a machine. He wrote:

> This will not seem strange to those, who, knowing how different *automata* or moving machines can be made by the industry of man, without employing in so doing more than a very few parts in comparison with the great multitude of bones, muscles, nerves, arteries, veins, or other parts that are found in the body of each animal. From this aspect the body is regarded as a machine which, having been made by the hand of God, is incomparably better arranged, and possesses in itself movements which are much more admirable than any of those which can be invented by man.[22]

In a later work Descartes argued that life was to death as a watch wound up was to one that had run down.[23]

Descartes developed an entire physiological system based on mechanical principles, and, in the "Discourse on the Method," he interpreted Harvey's circulation as a hydraulic system that functioned according to the laws of mechanics.[24] His model of the human body was influenced by the mechanical and hydraulic statues that he saw in the royal gardens at Saint-Germaine-en-Laye.[25] In a suppressed work, *Treatise of Man* (written between 1629 and 1632), Descartes described mechanical physiology. He argued that some parts of the blood penetrate into the brain where they serve as a source of the fluid which activates the body. From the brain the fluid enters the nerves, which Descartes viewed as small tubes, where it activates the muscles. He noted that,

> as you may have seen in the grottoes and fountains in our gardens, the force with which the water issues from its reservoir is sufficient to put into motion various machines, and even to make them play several instruments, or pronounce words, according to the varied disposition of the tubes which conduct the water. Indeed, the nerves of the machine may very well be compared with the tubes of these water-works; its muscles and tendons with the other various engines and springs which seem to move these machines; its animal spirits to the water which impels them, of which the heart is the source or fountain; while the cavities of the brain are the central reservoir.[26]

For Descartes the human body could be completely explained as an automaton, but human beings were not simply bodies. The prime quality of humans was the rational soul which, being spiritual, could not be explained by the mechanical properties of matter, so that human beings were more than simply automata. Descartes was never clear on the exact relationship between the soul and the body. At one point he held to the older view that the soul was located in the pineal gland, which could not be found in animals and therefore was associated with human rationality.

Although the presence of a soul prevented humans from being reduced to automata, animals were simply *bête-machines* (beast machines).[27] According to Cartesian philosophy, reason must be spiritual and thus immortal. If an animal possessed any reason it would also have to be granted an immortal soul and, for Descartes, such a belief would threaten the primacy of the human soul. Therefore, Descartes theorized that an automaton built with enough skill would be indistinguishable from the animal itself. He was forced to admit that animals experienced sensations, but unlike those of humans, which involved a spiritual process, these sensations were purely materialistic. For Descartes, the animal soul was not spiritual but simply blood, from which a rarefied fluid derived and served as a reservoir to activate the nervous system.

Descartes' denial of the animal soul raised a great deal of debate. Some people accepted his position that the existence of an animal soul would threaten the authority of scriptures. Descartes' position gained stronger support from the occasionalist philosophy of Father Malebranche, who believed that there was no actual connection between the body and the soul; rather they were simply the "occasion" for each other.[28] Only through God's constant intervention could one affect the other, so the fact that animals appeared to act with intelligence, while lacking a soul, simply proved the great providence of God.

Many people feared that denying the animal soul was more of a threat to religion than accepting it.[29] They supported the idea that animals could have souls which allowed a certain level of intelligence but not rationality. Basically they wanted to deny that life in any form could be reduced to pure mechanisms. Admitting that animals were *bête-machines* opened the door for extension of the concept to humans. As we shall see, by the middle of the eighteenth century, the final step was taken to identify man as a machine. Unlike Harvey, Descartes was convinced that physiology could be explained in terms of mechanical philosophy. In doing so, Descartes provided a philosophical basis for assuming that organic life was governed by the laws of mechanics.

Iatromechanists

During the second half of the seventeenth century, the mechanical world view began to have an influence on some of the physiologists themselves. Although, as we shall see in later chapters, there were other theories of life, the so-called iatromechanical (*iatro* meaning medical) movement had a significant impact on the development of physiology. Many of the iatromechanists extended Harvey's quantitative approach to other biological processes.

An important group of Italian iatromechanists was particularly influenced by Galileo.[30] Galileo spent most of his time investigating the inorganic physical world, but his distinction between primary and secondary qualities had a direct effect on physiology. The fact that tastes, smells, and temperature (secondary qualities) could be explained by the geometrical properties of matter (primary qualities) implied that the sense organs functioned according to the laws of mechanics. For example, Galileo explained taste as follows: "Particles . . . land on the upper part of the tongue, mix with its moisture and penetrate its surface, and cause the tastes, sweet or unpleasant according to the differences in the way their several shapes touch and according to whether they are few or many, and slow or fast."[31]

One of the earliest iatromechanists was the physiologist and mathematician Giovanni Borelli. He studied with Benedetto Castelli, who had been a student of Galileo, and with Descartes.[32] While at the University of Pisa, Borelli established an anatomical laboratory in his own home, which served as a training center for the iatromechanical movement. It was here that Borelli began the research for his major physiological work, *On the Motion of Animals,* which was not published until the year after he died. In the dedication he stated that God, being a geometer, had used geometry to construct the world.[33] Since the action of animals was essentially that of bodies in motion, it too must be subject to geometrical and mechanical principles.

Much of Borelli's work was an attempt to explain muscular motion in terms of mechanics. He treated the bones as if they were a series of levers, and reasoned that, because the muscles were attached close to the joints, or fulcrums, they had to overcome a great mechanical disadvantage to move an arm or leg. Through a series of erroneous calculations, Borelli concluded that the muscles would have to develop a tension of several thousand pounds in order to lift a relatively light object.[34] This error showed one of the weaknesses of the iatromechanical approach. The body could be treated as a machine in a qualitative sense, but it was always much more difficult to make exact numerical calculations.

Borelli was also concerned with the actual motions of the muscles and their relationships to the nervous system. He believed that the nerves, like the veins and arteries, carried some type of material or fluid substance. Muscular motion was brought about when the nervous system injected some of the nervous fluid into the muscle. When this happened, "something like a fermentation or ebullation" took place, "by which the sudden inflation of the muscle is brought about."[35] That is, the nervous fluid caused a chemical reaction in the muscle which made it inflate and move.

Borelli also argued that the brain served as a direct material connection between the sensory nerves and the nerves activating the muscles. If a nerve was stimulated by heat or some type of pain, the nervous fluid would travel up the nerve to the brain, stimulating it and causing the brain to send some nervous fluid to the appropriate muscle. Because of such elaborate explanations, Borelli's *On the Motion of Animals* became the foundation stone in the iatromechanical movement.

One of his students, Marcello Malpighi, continued Borelli's work and made several important discoveries in the field of physiology.[36] Malpighi studied medicine at the University of Bologna and then went to Pisa where he came under the influence of Borelli. While at Pisa, Malpighi participated in some dissections in Borelli's laboratory where he became converted to mechanical–atomistic philosophy. He eventually left Pisa to teach at Messina and later at Bologna, finally becoming the chief physician to Pope Innocent XII.

Malpighi sent his first work, *On the Lungs* (1661), to Borelli. According to traditional physiology, the lungs were thought to be hot, moist, fleshy organs with little structure. Malpighi used a series of lenses with varying magnifying powers to study the lungs of a frog. He was using this optical technique, which derived its inspiration from Galileo's work with the telescope, even before Atonj van Leeuwenhoek "invented" the microscope.[37] What Malpighi observed was that the lungs had a definite structure, composed of air cells surrounded by a network of small blood vessels—the capillaries. This discovery was extremely important since it provided the last link in the theory of the circulation of the blood. Without the microscope, Harvey had been unable to find the required connection between the veins and the arteries, but Malpighi's study of the lungs provided empirical evidence that the capillaries existed.

During his years at Messina (1662–66), Malpighi extended his iatromechanical researches to other parts of the body. In his study, *On the Tongue* (1665), Malpighi discovered that the tongue contained a series of pores which led to one set of sensory receptors—the papillae.[38] This work provided proof of Galileo's speculations concerning taste.

Malpighi's combined interests in atomism and iatromechanics also led to a new theory of the glands.[39] He believed that the glands were secretion machines whose function was to separate particles from the blood and send them, as a new fluid, into a secretory duct. Although he could not observe the structure of the glands with his microscope, he postulated that they functioned mechanically. Malpighi believed that the glands were essentially sieves. Located at the junction between a vein and the secretory duct, a gland, because of the size and shape of its pores, would allow only certain particles to pass from the blood vessels into its ducts.

Malpighi made microscopic studies of plants in the later years of his life, and found that they too had a mechanical structure.[40] In an essay written at the end of his life, Malpighi argued that the mechanisms of our bodies were "composed of strings, thread, beams, levers, cloth, flowing fluids, cisterns, ducts, filters, sieves, and other similar mechanisms."[41] He believed we could understand the function of organisms from a study of mechanical devices.

During the seventeenth century mechanical philosophy was also applied to embryology.[42] Before the 1670s, the most widely accepted idea of reproduction and development was epigenesis—the belief that an animal or human embryo developed from an egg composed of undifferentiated matter which then gained some form during fertilization. In his work *De la formation de l'animal* (1664), Descartes tried to explain epigenesis in terms of mechanical philosophy by assuming the existence of male and female semen, each of which contained a variety of particles. During fertilization these particles came together and through their motions began forming the various organs of the embryo. Descartes' explanation of epigenesis did not attract a wide following since it seemed unable to reveal how the motion of particles produced a series of organs perfectly arranged to form an organism.

The problems with epigenesis led to the rise of a rival theory of embryology. In 1674, Nicolas Malebranche proposed a theory of preformation based on the concept of *emboîtement* or encasement. Preformation held that the embryo already existed in some form encased in the egg (or according to some, in the sperm), and it simply needed to be allowed to grow in size. Inside each preformed embryo would be, like nesting Russian dolls, one set of even smaller embryos so that Eve's eggs contained all of the future generations. Several physiologists, such as Jan Swammerdam, claimed to observe completely formed animals in eggs and sperm.[43]

In the second half of the eighteenth century, one of the most vocal supporters of preformation was the Swiss physiologist Albrecht von Haller, who carried on a long debate with the German physician Caspar Friedrich Wolff, a supporter of the theory of epigenesis.[44] Through experiments on chicken eggs, Haller developed a mechanical theory in which fertilization caused the embryo's preformed heart to begin beating and to send fluids throughout the other organs so that they could begin growing and become observable. This theory could also explain one particular objection to preformation: Why did animals have characteristics resembling not only their mother but also their father? Haller argued that during conception the male semen could cause more fluid to be sent to some parts of the embryo than others, causing those parts to resemble the father more than the mother. Haller's physiology modified the mechanistic approach to life. As historian Shirley Roe has shown, Haller believed in mechanical philosophy but not in complete reductionism.[45] Haller argued that the laws functioning in living organisms were not necessarily the same laws that functioned in the inorganic world, but both sets of laws operated according to mechanical principles.

The mechanistic approach to physiology was further developed and transformed in the eighteenth century as a result of the success of Newtonian science. Much of physiology became dominated by the idea of a dynamic corpuscularity—that is, attractive and repulsive forces acting between particles rather than simply matter and motion.[46] The founder of this Newtonian school of physiologists was Archibald Pitcairne, who studied at Edinburgh and Paris, and later at Rheims, where he received his M.D. He then practiced medicine at the University of Edinburgh and the University of Leyden.

Pitcairne based his physiology on Newtonian principles. He believed that disease was caused by changes in the velocity or texture of the blood arising from forces acting between the particles. It is interesting to note that this theory gave new justification for the continued use of bleeding to cure disease. Whereas, in the Middle Ages, bleeding was done to eliminate evil spirits from the body, it was now done to "correct" the velocity of the blood and thus eliminate the disease.

Another Newtonian physiologist, John Freind, offered an ingenious theory to explain the menstrual cycle.[47] Earlier studies, comparing food intake and excrements, had found more material was taken into the body than was expelled. To explain the difference, physiologists postulated that there must be some "insensible perspiration," derived from the blood, which is given off in order to maintain a balanced weight. Freind argued that women had finer pores and weaker hearts, so their insensible perspiration must be less than men's. This difference caused fluid to build up to the point where it would break through the uterine vessels, which, being bent and twisted, were less resistant to bursting than other vessels.

There were also attempts to explain muscular action by Newtonian principles. A leading physiologist, James Keill, argued that particles of air penetrated the lungs and mixed with the blood and became highly compressed because they were surrounded by mutually attracting particles of blood.[48] Movement would take place when nervous fluid was injected into the muscle. Based on the Newtonian doctrine that small particles could get closer together and therefore were more attractive, Keill postulated that the small particles of nervous fluid would attract particles of blood from the air, allowing the air in the muscle to rapidly expand and activate some motion.

The high-point of British iatromechanists was the work of Stephen Hales.[49] Educated at Cambridge, he became chaplain to George, Prince of Wales, later King George III. Although he had little training in physiology, his *Haemastaticks* (1733) became one of the major works in iatromechanism. In this work Hales carried out a series of experiments to measure the various parameters of iatromechanical physiology. Some of these experiments, carried out on living animals, were gruesome. For example, he attached tubes to the arteries and veins of horses and dogs to measure their blood pressure, finding that it was not high enough to explain muscular action. Also by forcing water through the veins and arteries to measure their strength, he found that Freind's hypothesis of bursting uterine vessels must be false. Although his experiments cast doubt on whether some iatromechanical theories were quantitatively correct, they were done in the spirit of iatromechanical physiology.

The Cartesian and Newtonian view of physiology was combined in the work of Herman Boerhaave.[50] He studied both theology and medicine in his native Holland, where he heard the lectures of Pitcairne. Boerhaave spent most of his life at the University of Leyden where he taught medicine, chemistry, botany, and mechanics. His work was well known throughout Europe, especially England, and many of his students became the leading scientists of the eighteenth century.

Like Malpighi, one of the strengths of Boerhaave's physiology was that he combined his theory with empirical observations. Concerning the steps of his method, he said, "The first is an accurate *observation* of all the appearances offered to our senses . . . the *second* is a strict consideration and discovery of the several latent causes, concealed from our *senses* in human bodies."[51] Not all the "latent causes" of an organism were necessarily mechanical; Boerhaave did not exclude ideas such as "vital force" or "animal powers" from his physiology, although he never discussed them in any great detail. He seems to have accepted the Cartesian dualism between mind and body. For him, mechanical theories could not explain "memory, understanding, reason, and the knowledge of past and future appearances."[52] Although he could not completely reduce the living organism to a machine, Boerhaave argued that "if some portions of the human body correspond in their structure with mechanical instruments, they must be governed by the same laws."[53] According to him we find some organs of the body "resembling *pillars, props, crossbeams, fences, coverings;* some like *axes, wedges, levers,* and *pullies;* others like *cords, presses,* or *bellows;* and others again like *sieves, strains, pipes, conduits,* and *receivers.*"[54]

Boerhaave, like Pitcairne, viewed the body as a hydraulic machine. Not only did the blood circulate but so did the other fluids including the lymphatic fluid and nervous juices. The muscles were composed of a series of fibers that would act when the blood exerted a pressure in the space between the fibers. Digestion could also be explained in terms of mechanical processes. Boerhaave theorized that the primary action of digestion was the wearing away of solids by abrasion, first in the mouth where the teeth cut and ground the food, and continuing in the stomach, where various fluids caused an "intestine motion," or fermentation, of the food particles. Once the food was broken down into small particles it would be carried by the blood to the parts of the body where the food replaced those elements that were wasted away by normal use.[55]

The Clockwork Human

For Boerhaave, as for many of the iatromechanists, the machine was no longer just an analogue for organic processes—it described an actual reality. The functions of the human body, such as circulation, respiration, digestion, sensation, and movement, were actual mechanical processes. Of course, there was still the question of the mind or the rational soul. The Cartesian dualism of mind and spirit provided some basis for the belief that the body was subservient to the nonmaterial soul, but Descartes was never clear about the actual relationship between mind and body. As iatromechanical theories became more and more powerful and self-sufficient, it became less and less clear that the mechanical processes of the body depended on a nonmaterial cause. As previously discussed, Descartes himself believed that animals—*bête-machines*—could be described entirely in terms of mechanical causes. Once animals were thought of in this way, it was a short step before humans too were thought of as machines.

That step was finally taken by Julien Offray de la Mettrie, who brought about further modifications in a mechanical physiology.[56] Born in France, he studied philosophy and natural science at the College d'Harcourt, one of the first French institutions to teach Descartes' works after they had been banned by the Church. La Mettrie then studied medicine in Paris and at Rheims, but realizing the shortcomings of medical education in France, he went to Leyden where he studied with Boerhaave, before returning to practice medicine in France. La Mettrie was greatly influenced by Boerhaave and disseminated his works in France. Since La Mettrie was critical of both medicine and the Church in many of his early writings, he found it necessary to leave France and live in Holland. There he wrote his most famous work—*L'Homme machine* (*Man a Machine*, 1748). The book challenged the assumptions of medicine, philosophy, and theology by supporting a materialistic position—a belief that life could be explained by material causes. This publication caused such an uproar that La Mettrie was forced to flee to Berlin, where he came under the patronage of King Frederick II. La Mettrie died in Prussia after indulging in a great quantity of *paté*—a death that his enemies saw as entirely fitting for the great advocate of materialism.

L'Homme machine was a philosophical work with roots in Cartesian mechanical philosophy. La Mettrie's ideas differed from Descartes' in three fundamental ways.[57] Unlike Descartes, La Mettrie argued that the soul was inseparable from the body; the general problem of understanding the nature of the mind was a problem of physics. Second, La Mettrie denied that the concept of the beast-machine could not be extended to humans. If animals could be described in terms of automata, so could humans. Third, Descartes' biology as well as his philosophy were dependent on the existence of God to bring about interaction between the mind and the body, however, La Mettrie was an agnostic whose theory did not need an active God.

In *L'Homme machine,* La Mettrie tried to show that life, in all of its manifestations, could be based on a single cause—the principle of movement. Like his teacher Boerhaave, La Mettrie believed that the body, especially the nerves and muscles, was composed of a series of fibers. He was also influenced by the work of Albrecht von Haller, another student of Boerhaave's, who had attributed the property of irritability to the muscle fibers.[58] Knowing that muscles completely cut off from the body could be artificially stimulated into some type of motion, Haller postulated that the muscle fibers had an inherent property, irritability, that allowed the muscle fiber to move independent from the body.

La Mettrie transformed the mechanical idea of material by extending Haller's idea of muscular irritability to all matter, so that, along with such properties as weight and extension, all matter had the inherent property of movement. Given this redefinition of matter, La Mettrie was able to overcome the old Cartesian dualism between body and soul. Since one of the properties of matter is movement, La Mettrie no longer needed to appeal to the idea of a soul as the source of all vital activity. The soul emerges simply from the organization of the motile material that made up the body.

Although La Mettrie was far from clear in the presentation of his theory, he appeared to argue along the following lines.[59] The fiberous structure of the body is composed of material that has the property of inherent motion, with each fiber having a characteristic vibration. When the fibers are arranged to form various parts and organs of the body, the sum of all of the vibrations of each fiber results in the characteristic action of that part or organ. For example, the ability of a muscle to contract depends on the vibration of the muscle fibers, although the actual contraction occurs only with arrival of fluid from the nervous system. In a similar way, the mind itself could be explained by several levels of vibrations—at the lowest level, sensations cause specific vibrations in the brain, while at higher levels, other vibrations permit the sensations to be ordered, compared, and retained. As proof of the mind's dependence on the body, La Mettrie cited the effects on the mind of such obviously material substances as opium, wine, and coffee.[60]

By rejecting Cartesian dualism and redefining matter, La Mettrie established a material basis for life. As with Descartes, all matter had the property of extension, but, like Newton who added the active property of force to matter, La Mettrie gave matter the active property of inherent motion so that if it is organized in a particular way, it can also give rise to cognition. The more complex the organization of matter, the higher the mental functions; the difference between humans and animals was simply one of organization.

For La Mettrie, humans had become machines. At the same time, in France, there appeared elaborate attempts at automata in the forms of animals and humans.[61] Some of the most famous works were the creations of Jacques Vaucanson, who was a great influence on La Mettrie. Born in France, Vaucanson studied for the Church but was relieved of his vows and had his workshop destroyed after he was said to have created angels that flew mechanically. He then moved to Paris where he developed interests in medicine and music, while retaining his interest in mechanics.

On his arrival in Paris in 1735, Vaucanson began work on three automata that would make him famous. The first was a life-sized figure of a musician, which, through a series of bellows and levers, played eleven songs on a flute, blowing into the instrument with its mouth and moving its fingers over the levers on the flute. The statue was so detailed that the lips and the tongue functioned similarly to those of a real musician.[62] His second automaton was similar except that while playing a drum with one hand, it played a provincial flute with the other.[63]

Vaucanson's most astonishing work, however, was his third automaton, a duck that he exhibited in 1738. Rather than simply recreating movements, he attempted to reproduce the major organic functions of a duck. According to Vaucanson, the automaton was an animal "made of gilded copper who drinks, eats, quacks, splashes about on the water, and digests his food like a living duck."[64] The work was exceedingly complex; one wing had more than four hundred pieces. Run by a complicated set of gears and tubes, many of which were concealed in the base of the work, the duck would swallow grain that was presented to it; the grain was then digested inside, through a system of mechanisms, and eventually excreted. Vaucanson even provided a section through which the digestive process could be observed. There is evidence that, with the monetary support of Louis XV,

Vaucanson also attempted to create a complete artificial man, but he died before the project could be completed.[65] Vaucanson's automata were famous throughout Europe. Even after his death they were displayed in Russia and in Germany, where they were seen by Goethe.[66]

During the last half of the eighteenth century, automata became more and more complex and were able to imitate many human actions. Pierre Jacquet-Droz and his son Henry-Louis created automata that could write or draw figures. The *Writer*, completed by 1772, was a small boy seated at a bench. Through a complex clockwork system the boy would dip his pen into the inkwell and begin to write a preprogramed sentence. He would distinguish between light and heavy strokes, would leave spaces between words, and would move to the next lower line when required.[67] The *Draughtsman* worked in a similar fashion except that, using a pencil, it would blow the dust from the picture when it was through.[68]

It seems clear that the makers of the automata had a higher goal than simply creating toys to amuse the royal courts of Europe.[69] They showed that the iatromechanical theories actually could be demonstrated. Technology was coming very close to recreating organic processes. There was, of course, the problem of the mind, especially the rational mind. But La Mettrie had postulated a material basis for the mind, and the automata of Jacquet-Droz seemed to be able to imitate some of the rational aspects of humans.

The Mechanical Mind

The closest thing to a mechanical theory of the mind was put forward by John Locke in his *Essay Concerning Human Understanding* (1690).[70] Although he is best known as a philosopher, Locke studied medicine at Oxford where he became close friends with Robert Boyle, whose corpuscular philosophy had a strong influence on him. In the *Essay*, Locke argued for what could be called a corpuscular theory of ideas. For him, the main source of ideas was sensations that arose from external matter, impinging on our sense organs. These sensations then resulted in simple ideas in the mind that could be combined into complex ideas, not unlike the ability of primary qualities of matter and motion to produce secondary qualities such as color, smell, tastes, and sounds. Otto Mayr has shown that throughout the *Essay*, Locke explained the relationship of the internal constitution of the mind to the observed faculties of sensation and reasoning in terms of the internal gears and springs of a clock producing the external motion of the hands.[71]

In the eighteenth century the Lockean theory of mind was extended in the associationalist psychology of David Hartley.[72] Hartley, a friend of Stephen Hales, was educated at Cambridge and later turned to medicine, although he never received a formal medical degree. Most of Hartley's theory of mind is contained in his *Observations on Man* (1749). His works were very popular in England during the second half of the eighteenth century, and had such an impact on Samuel Coleridge that he named a son Hartley.

Hartley attempted to apply a materialistic idea of Newtonian mechanics to a theory of the mind. The basis of his theory was the belief that sensations of the external world impressed small vibrations on the material of the nerves. These vibrations were then transmitted to the brain where they caused a particular vibration in the cerebral material. Since these sensations could produce vibrations of different kinds and strengths and in different locations, each vibration represented a simple or unit idea. After repeated sensations of a certain type, that part of the brain would become predisposed to vibrate in a given way. If two sensations arrived in the brain simultaneously they would become "associated" with one another. In such a case, the vibrations would modify one another, representing complex ideas, or the vibration of one area of the brain could cause the other area to vibrate even without any external stimuli—representing memories or recollections similar to when a certain sound or smell would cause us to recall visions of our past.

As M. H. Abrams has shown, Hartley's associationalist psychology led to a mechanical theory of creativity.[73] Ideas in the mind simply represented a mirror-image of the external world. And, as Hartley had shown, the creative function of the mind consisted of breaking down complex ideas into their component parts and rearranging them into new ideas. That is, the imagination functioned in a purely mechanical way—simply rearranging parts into new wholes.

Conclusions

By the end of the eighteenth century, mechanical philosophy seemed able to explain organic life, from the function of bodily organs to the creative aspects of the mind. Such mechanical theories of life play an important role in understanding the relationship between technology and organic life. Within the mechanical world view it was possible to reconcile the basic differences between machines and living organisms. Even though organisms appear very different from machines, as did the shadows on the wall of Plato's cave from the objects that caused them the supporters of the mechanical world view would argue that we should not be deceived. In the reductionist world of mechanical philosophy, machines and organisms could both be explained in terms of mechanical principles. The apparent conflict between the two is resolved by reducing life to technology. Life in general, even human beings, were at their base functioning as mechanical organisms.

But some problems with the mechanical theory of life remained. This theory found difficulty with the relationship between the active nature of the mind and the passive quality of the material of the body. Descartes' dualism separated the realms of mind and body, but it was still not clear that mind could be reduced to matter. Even the mechanical theories of mind and soul, such as La Mettrie's and Hartley's, required a redefinition of matter so that it could be endowed with active properties such as motion. Organisms were machines but the material of these machines was no longer inert and passive.

Through the development and modification of the idea of mechanical philosophy, the mechanistic theory of life had come much closer to the opposing organic world view. During the nineteenth century, developments in areas such as electromagnetism and thermodynamics could no longer be simply or easily reduced to Newtonian mechanisms.[74] That is, while the organic world was becoming mechanical, the technological world was moving toward a different philosophy—an organic world view.

4

The Organic World View:
From Magic to Vitalism

Sitting quietly, doing nothing,
Spring comes, and the grass grows by itself.
                                A Zen poem

The mechanical philosophers conceived of the world as a giant machine, and they believed that the objects in the world, even living organisms, were smaller mechanical devices. Natural processes, including life itself, were simply the result of the mechanical motions of material bodies. Some of the early mechanical philosophers, such as Descartes, believed in a separate and distinct spiritual world that exerted some guiding influence on the material world. But philosophers had found it difficult to explain the exact connection between the two worlds, usually leaving it up to God to establish some harmony between them. As more and more processes could be explained through matter and motion, philosophers and scientists, especially in France, began to abandon the spiritual world. While the mechanical world view had dominated the science and philosophy of France, England, and Holland throughout the sixteenth, seventeenth, and eighteenth centuries, during the same period another world view attracted the attention of some scientists and philosophers, especially in the Germanic areas of Europe.

What could be called an organic world view encompassed a wide range of philosophies, including among others, magic, alchemy, Hermeticism, vitalism and Romanticism.[1] Over time, this world view has undergone significant development and change, usually in response to the challenges of the mechanical world view. But some basic characteristics have remained consistent. Rather than reducing phenomena to their simplest constituent parts and analyzing those parts in isolation from the whole, the organic world view was antireductionist and emphasized the organization or interrelationship of the parts. The whole manifested new purposive behavior that could not be accounted for by the actions of individual parts. This goal-seeking or teleological behavior implied the existence of some directive agency, such as vital substances, vital forces, or vital principles,

46

in addition to the mechanical concepts of matter and motion. Such spiritual or vital principles not only determined the form and development of biological organisms, but could also permeate the entire universe so that a wide range of objects beyond biological organisms were thought to be alive to some degree. While the mechanical world was thought to function like a machine or clock, this world was thought to function like a giant organism or plant.

Some of the elements that contributed to the organic world view appeared as far back as the philosophy of Plato, Aristotle, and the Stoics.[2] As we have seen, Plato's theory of ideal forms played an important role in the development of mechanical philosophy, but his theory of nature also had an organic element that supported a different view of the world. In the *Timaeus,* Plato discussed the structure of the universe. As in his other works, he argued that the universe was based on ideal mathematical principles, the various elements of the world, such as air, water, earth, and fire, composed of so-called Platonic solids (solids with regular figures on each face, for example, a cube or a tetrahedron).[3] But the universe, besides being mathematical, possessed another quality. Not only did it move in circular motion by itself, the universe was rational; hence a product of some divine intellect or World Soul.[4] To Plato the visible world was an image of the eternal world of Ideas or Forms. But this visible world was connected to the ideal world through spiritual entities which Plato called "demi-urges." These spirits were the creative powers in the world, imposing an order, drawn from the model of eternal ideas, onto the chaotic matter of the visible world. Even though Plato's universe was rational and mathematical, giving support to a mechanical world view, its very rationality was a manifestation of a nonmechanical World Soul, the basis of an organic, mystical world view.

Plato focused most of his attention on the world of ideal forms, using mathematics as his model for understanding the world. His successor, Aristotle, approached the study of philosophy in another way. Aristotle, whose background was more biological than mathematical, focused his attention on the problem of describing change in the natural world rather than the problem of describing the eternal forms of the ideal world. Over one third of Aristotle's work was in the field of biology, leading some scholars to argue that his thought should be approached through biology rather than philosophy.[5]

Aristotle inverted the relationship between the organic and the inorganic. Instead of explaining organic processes in terms of inorganic concepts as the mechanical philosophers would do, Aristotle assumed that organic development was fundamental; he explained material changes in terms of organic models.[6] Aristotle was always concerned with the question of change. Unlike later philosophers who considered matter or substance to be unchanged or constant throughout a process, Aristotle thought that substance was different at the end of a process from what it was at the beginning.[7] All matter, like the material of a seed, had the potential to develop distinctively into something else. For Aristotle even the motion of an object from one place to another was analogous to the change of an acorn into an oak tree. That is, unlike purely mechanical motion which required some outside cause, motion for Aristotle was inherent in the object being moved, just as a mature plant is inherent in a seed. An outside agent, such as a person throwing a ball, may in Aristotle's words actualize the motion, but its source is still in the object itself. As the philosopher John Herman Randall has noted, "He takes biological examples, living processes, as revealing most fully and clearly what natural processes are like. He analyzes the behavior of eggs, not of billiard balls."[8]

After the death of Aristotle in 322 B.C., Greek philosophy came to include more and more mystical elements. This can be attributed partly to the influence of other philosophies as the center of Greek culture shifted from Athens to Alexandria. One such philosophy was Stoicism; its influence began about 300 B.C. and persisted for over five centuries, especially in imperial Rome, where Emperor Marcus Aurelius was himself a leading Stoic philosopher. The basic belief of the Stoics was determinism. Everything in the world happened according to some law; human beings must try to understand these laws in order to live in harmony with them. Although this philosophy might have led to a rational mechanical world view, for the Stoics it led to mystical beliefs in such things as astrological divination based on harmony between celestial and terrestrial events.[9]

The Stoic belief in a universal harmony led to a view of nature that was radically different from the more mechanical view of the atomists. The atomists, such as Democritus and Epicurus, had believed that the order of the world occurred only by chance, through the random interaction of a large group of independent atoms. Living organisms were simply atoms that came together by chance for a period and then, again by chance, moved apart. But the Stoics believed that any organized system must be more than simply the sum of its individual parts. Something else must exist to give the system its order and its properties as a whole. This something was the *pneuma*. It was thought to be a special material, extremely tenuous, like a gas, which was continuously spread throughout the parts of an organized system such as an organism. Since the Greeks believed there were four elements, earth, water, air, and fire, the *pneuma* became associated with a mixture of air and fire which then would hold together bodies composed of earth and water.

The *pneuma* provided the basis for a harmony between all of the elements in the world. Since *pneuma* was used to explain the properties of organisms, it became associated with the basic principle of life, what the Greeks called *psyche,* which later was translated as soul. But *pneuma* was present in other things besides animate objects. The universe as a whole was an ordered system, and so the Stoics believed that it must be bound together by a universal *pneuma*. This "World Soul," which many identified as the Deity, was thought to be the source of the individual *pneuma* of the various bodies in the universe. Galen, the physician to Marcus Aurelius, had postulated that the World Soul entered human beings through the lungs as they breathed. The universal *pneuma* was composed purely of fire, while the *pneuma* of actual bodies contained both air and fire. Therefore all objects in nature, both animate and inanimate, were alive, and the source of this life was the World Soul of the universe itself. Also because the World Soul was the source of the *pneuma* of both individuals and of the stars and planets, a connection between human beings and the rest of the universe was obvious.

Alchemy and Astrology

This relationship between the entire universe (the macrocosm) and a small part of it (the microcosm), such as an individual, can best be seen in the development of alchemy and astrology. Although most people, especially scientists, in the modern world give little credence to these theories, historians now realize that they played an important role in the development of modern science.[10] Both of these theories were connected to a theological interpretation of science that was based on an organic or vitalistic view of the world.

The origins of alchemical theory are hidden in the ancient past. Most likely, it began as a fusion of ideas from the philosophies of both East and West.[11] Although today the term alchemist suggests a mad scientist driven by greed to an obsession with turning base metals into gold,[12] alchemy was closely associated with the theological desire to gain more direct knowledge of the Creator. For this reason the alchemist placed great emphasis on direct observation and experience.

Alchemy was also strongly dependent on a belief in the macrocosm–microcosm analogy. The alchemist believed that when he conducted an experiment on a small part of the universe, he was in fact conducting an experiment on the entire universe. Therefore, if the alchemist succeeded in transforming some base metal like lead or iron into a more perfect metal like gold, he had actually acted on the entire universe and in so doing had himself undergone a transformation that would be similar, in many ways, to a religious rite. It should be noted that the belief that an experiment conducted on the microcosm could be seen as an experiment on the macrocosm has become a fundamental assumption of modern science.

Much of alchemy was focused on the problem of the Creation, especially the origin and development of the elements of matter that made up the world.[13] Inherent in much of Greek philosophy, especially the works of Aristotle, was the belief in some fundamental substance, or *prima materia,* out of which arose the four elements. The four elements originated from the *prima materia* when it was acted on by a pair of the qualities wetness, dryness, heat, and cold. Greek philosophy therefore provided some justification for the belief that the elements were created through a process similar to chemistry or alchemy. Also, since all the elements had a common origin in the *prima materia,* it might be possible to transmute one substance into another by removing one set of qualities, thus reducing the element to *prima materia,* and then substituting another set of qualities.

The alchemical theories of the elements were based on a vitalistic or organic view of nature. It was a common belief among metal craftsmen, blacksmiths, miners, and metallurgists that metals were alive and grew in the earth in a way similar to an embryo within a womb or an egg.[14] The World was viewed as *terra mater* or Earth Mother; everything within the earth was alive and in a state of gestation. As with an embryo, the metals went through distinct stages. The baser or more corruptible metals such as iron were thought to be in the very early stages of development, while more precious metals like silver were near the end of development, reaching their culmination in gold. This natural, organic transmutation seemed verified by the fact that metals were found in "veins" and by the frequent discovery of several different metallic ores within a single mine. If metals developed naturally within the earth, the role of the alchemist was simply to accelerate the processes of nature. The alchemist's crucible or furnace played the role of a surrogate womb or egg which imitated the Earth Mother or womb.

Alchemical theory encompassed many elements of astrology. In keeping with the microcosm–macrocosm analogy, seven metals corresponded to the seven known planets (Sun-gold, Mercury-metallic mercury, Venus-copper, Moon-silver, Mars-iron, Jupiter-tin, Saturn-lead), and the alchemists denoted the metals with the same symbol used to designate the planet.[15] Just as the planets were thought to control different aspects of human existence, the alchemists believed that each planet influenced a metal during the process of development and perfection within the earth. Therefore, many alchemical procedures could take place only when the planets were in a certain configuration. Also, the connection between the metals and the planets implied that alchemical creation of a metal was, in fact, a reenactment of the Divine creation of the universe.

After the decline of the Roman Empire and with it Alexandria, many alchemical and other mystical works were temporarily lost to the West. During the early Middle Ages the center of study for the mystical sciences was Islam. Throughout the learned centers of the Near East, the major Greek works of science were translated and analyzed. Islamic scientists and philosophers accepted but also refined much of Greek alchemical theory. Although they accepted the Aristotelian element theory, these authors also developed a theory which argued that all metals were composed of a combination of idealized substances that resembled common sulfur and mercury.[16] This sulfur–mercury theory may derive from the Stoics' belief in active and passive "spirits" (sulfur being fiery acted as the active spirit while mercury being liquid acted as the passive spirit).[17]

Like the earlier alchemists, Islamic scholars continued to emphasize an organic or vitalistic view of nature. Islamic literature went even further than the Greek tradition, however, in placing special interest in the relationship between alchemy and medical chemistry.[18] This connection probably had its roots in India and China where there was an early belief in physical immortality and a search for elixirs of life. Using the microcosm–macrocosm analogy, Islamic alchemists believed that the transmutation of base metals into gold should be similar to the perfection of the person. A search began for chemical substances of medicinal value rather than the more traditional herbal cures.

The development of European alchemy was dependent on and reflected the theories developed in Islam. By the twelfth century, many works of classical Greece, which had been lost to Western Europe during the early part of the Middle Ages, were being rediscovered through contacts with Islamic culture. Although most of the interest in the Latin West was in translating and analyzing the philosophical and scientific works of Aristotle, an interest in the theories of alchemy is reflected in the works of such thinkers as Thomas Aquinas and Albertus Magnus.[19]

Hermeticism

Yet not until the Renaissance did alchemy and other mystical sciences begin to play a significant role in the development of modern science. The Renaissance was characterized by a new concept of the individual's relationship to the cosmos, with increasing criticism of medieval texts and of the authority of Aristotle.[20]

Although the Renaissance can be seen as the root of the Scientific Revolution, which led to the new astronomical system of Copernicus and eventually to the works of Galileo and Newton, the period was dominated by a mystical or animistic view of the world. Many historians have sought to explain the new Renaissance attitude by emphasizing the growing interest in Platonic or Neo-Platonic philosophy, especially in Florence with the rise of Marsilio Ficino's Platonic Academy, which influenced artists like Michelangelo. But recently Dame Frances Yates and other scholars have argued that the core of Renaissance thought should be labeled Hermetic rather than Platonic.[21] The label Hermetic comes from a body of works on alchemy, astrology, and magic called the *Corpus Hermeticum*. In the Renaissance these treatises were thought to be the work of Hermes Trismegistus (Thrice Great Hermes), an "Egyptian priest, contemporary with Moses, a Gentile prophet of Christianity" and the source of the ancient wisdom that became Greek philosophy.[22] This work was considered so important that Cosimo de' Medici asked Ficino to translate it into Latin (completed 1463) before he translated the works of Plato; Hermeticism is at the core of Renaissance Neo-Platonism.

The texts of the *Corpus Hermeticum* presented a new concept of the Creation and of the individual's relationship to the cosmos. One treatise, the *Pimander,* describes the creation of the world and the creation of man. Unlike the biblical account of Creation, it makes a direct and spiritual link between the earth and the heavens; man himself is given divine and magical powers over the world.[23] The "Hermetic Adam" has a new relationship to the universe. Through the performance of magico-religious acts this new Adam or magus can act and operate directly on the world. Through the use of magic he is in direct communion with the entire universe. Pico della Mirandola echoed this new attitude in his "Oration on the Dignity of Man" with the line from the Hermetic treatise *Asclepius*: "A great miracle . . . is man."

Hermeticism supported an organic or animistic world view. Ficino himself developed an elaborate theory of magic that combined Hermeticism, Neo-Platonism, and Pythagorean musical theory.[24] He believed that the universe was alive and contained a soul which through the stars and planets imprints forms on the sublunar world.[25] He wrote, "Undoubtedly the world lives and breathes, and we may absorb its breath (*spiritus*)."[26] Ficino believed in the Stoic idea of a world spirit (*spiritus mundi*) that filled the entire universe and provided a "channel of influence" between the individual (the microcosm) and the heavenly bodies (the macrocosm).[27] Since the earthly form might be imperfect because of corruptions in the material world, the goal of the magician, like that of the alchemist, was to purify and nourish the spirit of earthly things by attracting and absorbing the world spirit. By means of magical sympathies one could capture celestial or planetary spirit; certain plants, foods, metals, or even music contained or attracted the effluvia of specific planets.[28]

This animistic view of the universe played an important role in Renaissance thought. Even scientists and philosophers who appeared to be "modern" and "rational" in their scientific discoveries were influenced by Hermetic ideas. For example, many of those who supported the sun-centered universe of Copernicus did so for reasons associated with solar magic as much as for reasons of empirical evidence. Even Copernicus himself stated that

> in the center of all resides the Sun. Who, indeed, in this most magnificent temple would put the light in another, or better place than that one wherefrom it could at the same time illuminate the whole of it? Therefore it is not improperly that some people call it the lamp of the world, others its mind, others its ruler. Trismegistus [calls it] the visible God.[29]

Giordano Bruno, one of the earliest supporters of the Copernican system, did so because of his belief in an extreme form of religious Hermeticism.[30] He argued that the earth's motion around the sun proved that the earth was alive.

Closely tied to the organic world view was the new interest in the magnet. William Gilbert, whose work *De magnete* (1600) became the classic study of the properties of lodestones, also supported an animistic view of the world.[31] The magnet had frequently been used as an example of the action of occult sympathies. For Gilbert, the effects of the lodestone could explain the animistic force (*anima motrix*) present in the universe. Although he did not accept Copernicus's heliocentric system, Gilbert did believe that the earth rotated daily on its axis. He supported this belief with the argument that the earth was a giant lodestone and that lodestones would rotate by themselves because of their animistic force.

Gilbert's "magnetic philosophy" had an effect on astronomer Johannes Kepler. Kepler, best known for his three laws of planetary motion, completed most of his work at the court of Emperor Rudolf II in Prague, which was a center for mystical and alchemical thought. A supporter of the Copernican system, Kepler argued that the planets were held in their orbits by an *anima motrix,* similar to Gilbert's magnetism, which emanated from the sun. An analysis of this force led to Kepler's second law, which stated that a line from the sun to a planet would move through equal areas in equal times. Further study of the *anima motrix* led Kepler to argue that the planets did not move in circular orbits, as had been believed since antiquity, but in ellipses (Kepler's first law). The elliptical shape of the planets' orbits gave further support to an organic view of the universe since the shape of the solar system would resemble an egg.

Chemical Philosophy

Although Hermeticism played a role in the new astronomical theories of the universe, the mystical theories of the fifteenth and sixteenth centuries led to what historian Allen Debus has called a chemical philosophy, which is as important for understanding the development of modern science as the more widely studied mechanical philosophy.[32] Like mechanical philosophy, chemical philosophy was meant to replace the scholastic Aristotelian approach to nature. Rather than relying on an idealized mathematical system, however, it would depend on direct observations in nature and in the laboratory to discover and establish natural relationships between individuals and the world.

The source of much of chemical philosophy of the sixteenth and seventeenth centuries was the work of the Swiss physician Paracelsus.[33] Born near Zurich, Theophrastus Philippus Aureolus Bombastus von Hohenheim, later known as Paracelsus (meaning *greater than Celsus,* an early Roman physician), studied alchemy and worked as an apprentice in the mines before going off to study in many of the major universities of Europe. Although there is no proof, he may have received a medical degree at Ferrara. He did spend the rest of his life as a physician, gaining some reputation by treating the humanist Erasmus, but had to move frequently because he challenged the more established medical theories.

Paracelsus's work focused attention on chemistry as a model that could explain both biological organisms and physical phenomena. His chemistry was not reductionist, rather it was founded on the universal sympathies that were thought to exist between the microcosm and the macrocosm. Paracelsus treated the human body as a chemical system, however, unlike the mechanical philosophers, he did not believe that chemical processes functioned according to mechanical principles. Instead, he saw those processes as governed by vital spirits. For Paracelsus, the universe was filled with vital spirits or souls. Every object in the physical world was associated with a vital spirit that gave it a specific function or purpose. The sources of these vital spirits were the stars, each of which had an *astra,* or spirit, that influenced and directed some organ, plant, mineral, disease, or cure in the terrestrial world.

After the death of Paracelsus, his disciples began producing texts and commentaries on his works; by the end of the sixteenth century, a school of Paracelsians were developing a vitalistic chemical philosophy that challenged the older Aristotelian and Galenist views of natural philosophy and medicine. The Paracelsians, living in the age of the Reformation, wanted to replace the older Aristotelian and Galenist systems, which were thought to be heathen, with a Christian–Hermetic philosophy that would draw on divine revelation (the Bible) and divine creation (nature).[34] For this group, chemistry (or alchemy) became the chosen model for understanding nature since it combined a direct observation of the world and a vitalistic–religious framework. Both Paracelsus and his followers interpreted the Creation as a chemical or alchemical process. Through extraction, separation, sublimation, and distillation, God was able to produce light, dark, heaven, and earth out of the chaos.

The chemical account of Creation led the Paracelsians to the problem of the elements. In place of the ancient Aristotelian elements of earth, water, air, and fire, most Paracelsians used the three principles (*tria prima*) of idealized mercury, sulfur, and salt—a modification of the old alchemical mercury–sulfur theory but expanded to reflect the trinity of the body, mind, and soul.[35] Since the Aristotelian system had served for centuries as the basis of cosmology, natural philosophy, and medicine (through the four humors), the introduction of a new system of elements was implicitly an attack on the established framework of Scholastic science.

Chemical philosophy attempted to interpret all the processes of nature in terms of chemical or alchemical operations.[36] As in older alchemical beliefs, the Paracelsians gave heat or fire a central role in their natural philosophy. Paracelsus himself had argued, "If I should say by way of example that something cannot burn, then I also mean that it cannot live."[37] In the same way that the sun contained the "soul of the world," the earth contained a central fire that was the source of geological phenomena such as volcanoes, hot springs, and mountain streams. Through a mystical interaction between the earth and the heavens, similar to the sexual union of a man and a woman, metals were created within the earth and allowed to develop, nurtured by life-giving spirits originating in the solar rays.

Although chemical philosophy was applied to the entire universe, its vitalistic framework made it especially applicable to the study of medicine.[38] If the world was alive, all scientists were in fact physicians. The basis of this iatrochemical (medical chemistry) theory was the microcosm–macrocosm analogy—the belief that the individual was a small replica of the entire universe. Although the theory led to some peculiar beliefs such as the weapon salve cure (treatment of the weapon that created the wound rather than the wound itself), it also led to significant new theories about disease and its treatment. Paracelsians were among the first physicians to reject the old Galenic theory that diseases arose from an imbalance of the four bodily humors and that this imbalance could be corrected with herbal medicines. Paracelsians argued that disease was caused by external factors which could enter the body through food or the air and become localized in one organ. Just as metallic "seeds" grew into specific metals, "seeds" of disease grew within the body and interfered with the life force (or *archeus*) of the specific organ. Unlike subscribers to the older Galenic theory, who treated disease by treating the entire body by means of herbal cures, the iatrochemists argued that chemically prepared mineral medicines with their own life force must be used in order to offset the *archeus* of the specific disease. For example, iron salts might be administered to anemic patients because iron was associated with Mars, the red planet and the spirit of blood and iron.

One of the leading supporters of Paracelsian chemical philosophy was the Flemish physician J. B. van Helmont.[39] Through his attempt to find the vital principle that was responsible for individual life forms, van Helmont helped to transform the vital principle into a vital substance. His analysis of burning substances led him to introduce into chemistry the new concept of gas (from the term *gaesen,* to effervesce) which he identified with the vital spirit of a particular object.[40] A gas retained the properties of the material from which it originated and therefore could be seen as the bridge that linked spirit and matter. Alcohol (*spirituous*), he reasoned, can affect our minds because of its similarities to the vital spirit (*spiritual*). The material nature of this vital spirit could be seen in burning organic material such as wood or coal. Van Helmont noted that if coal was baked in a closed container, little if any weight was lost, but if the material was burned in the open so that the "wild spirit" could be released, only a little ash remained, indicating that most of the weight of the original coal had been a vital spirit which was released as burning gases.

Vitalism

During the eighteenth century chemical philosophy began to undergo modifications that focused attention on the concept of a vital substance, force, or principle.[41] In some cases these modifications resulted from the growing success of Newtonian philosophy but were interpreted in ways that seemed to support an organic world view rather than a mechanical one. The belief in the existence of some vital substance, force, or principle arose from the attempt to explain living organisms by means of chemistry. While alive, organisms seemed able to resist chemical decay, but after death they quickly decomposed into ordinary chemicals. In exploring this question, many physiologists postulated the existence of some vital substance, force, or principle that interacted with ordinary mechanical matter in such a way as to explain the properties of life.

One of the leading proponents of vitalism was the German chemist and physician Georg Ernst Stahl.[42] Although he supported a corpuscular theory of matter, he did not believe that living processes could be completely reduced to mechanical explanations. Mechanical ideas could explain some aspects of living organisms but could not explain their goal-directed activity such as growth, development, resisting decomposition, and striving toward some purpose. For Stahl, what distinguished living organisms from the nonliving was the *anima* (or soul), an immaterial agent that directed matter toward some goal or purpose and preserved this matter from normal corruption. This *anima* acted on material by way of purposeful motion which Stahl argued was not an intrinsic property of matter but something that required an external cause. By associating *anima* with motion, he believed his system could overcome the Cartesian question of how the immaterial mind could affect the material body.

By the middle of the eighteenth century, the distinctions between vitalism and mechanism were becoming less extreme. Many vitalists had given up a belief that living organisms contained a vital substance or soul. They argued instead that life was governed by some vital force similar or analogous to Newton's force of gravity.[43] Such ideas were particularly popular among those who tried to explain embryological development in terms of epigenesis rather than preformation.[44] For example, in 1745 Pierre-Louis Moreau de Maupertuis argued that particles, such as those composing the heart or legs, were attracted from each parent to one another to form the individual organs of the embryo. In 1749 George Louis Leclerc, Comte de Buffon, put forward another theory in his influential *Histoire naturelle*. He suggested that an embryo developed when seminal fluids from both the male and female were acted on by a ''penetrating force'' similar to gravity or magnetism, bringing various organic particles together so that they could be molded into a new embryo. A year later John Turberville Needham proposed a universal vegetative force consisting of an expansive force and a resistive force in equilibrium. The continual tensions and interactions between these two forces were responsible for all vital activity including embryological development. Finally, in 1759 Caspar Friedrich Wolff, who carried on the long debate with Albrecht von Haller over epigenesis, introduced the concept of a *vis essentialis* (essential force) to explain embryological development. According to Wolff, in organic materials the *vis essentialis* caused like materials to be attracted and unlike materials to repel each other. One part of the embryo which has solidified because of the attractive force will secrete unlike matter through repulsion, which will then solidify to form another part of the embryo.

By the end of the eighteenth century, several physiologists supported a theory of a living principle, an active power that when added to matter resulted in a living organism.[45] The leading exponent of a living principle was the British physiologist and surgeon John Hunter. After observing in autopsies that stomach juices, after death, began to digest the stomach itself, he postulated the existence of some vital principle which protected living matter from the chemical action of digestion.[46] He used this principle also to explain that animals have the ability to maintain a constant body temperature despite significant changes in the temperature of the environment. Like those who advocated a vital force, Hunter saw a close connection between the active power of organisms and the active power of matter to attract other matter. He argued that in an organism the living principle "is therefore essential to every part, and is as much a property of it as gravity is of every particle of matter composing the whole."[47]

By the beginning of the nineteenth century, it was becoming more and more popular to associate the vital principle with electricity.[48] While conducting dissections on frogs' legs during the mid-1780s, the Italian physiologist Luigi Galvani discovered that the disembodied legs could be made to twitch if stimulated by electricity. Even more startling, he found that the legs would also contract when they were pinned to an iron railing with brass hooks. Such a phenomenon would be explained today as the result of two dissimilar metals and a small amount of acid creating electrical current similar to an automobile battery. But Galvani was convinced that he had discovered an inherent "animal electricity" that was fundamental to living organisms.

By 1793 Alessandro Volta, an Italian physicist, had challenged Galvani's interpretation, by proposing a purely physical explanation which argued that the contact between two metals created electricity. A short time later Volta showed that electricity could be generated from a pile of alternating silver and zinc disks separated by moist cardboard. But many scientists, especially physiologists, continued to accept Galvani's association between electricity and the vital principle. Giovanni Aldini, a nephew of Galvani, conducted a wide range of highly dramatic experiments, such as causing the eyes and tongues of recently decapitated animals to move when electricity was applied to them.[49] His most highly publicized experiment was performed in London in 1803 when electricity was applied to the body of a recently executed murderer, causing the corpse to move. This convinced Aldini that galvanic electricity might be able to restore life to victims of recent drownings or suffocations. In other experiments, electric shocks applied to the chest caused a corpse to blow out a lighted candle placed in front of it. During an infamous experiment in 1818 on a recently executed criminal in

Glasgow, the muscular contractions caused by electricity became so lifelike that Andrew Ure, the well-known chemist conducting the experiment, had the corpse's jugular vein cut to insure it would remain dead.[50] These experiments along with the publication of Mary Shelley's *Frankenstein,* in which galvanism provided the "spark of life" that animated the monster, led to a widespread belief that electricity was the vital principle.

At first the association of the vital principle with electricity might seem like a step toward reductionism and a mechanical world view of life, but electricity was not a phenomenon that could be easily classified as mechanical. Throughout much of the nineteenth century it was an open question whether electricity was a material phenomenon or the result of some unknown force similar to gravity.[51] Also, during the nineteenth century it was difficult, if not impossible, to reduce electrical phenomena to mechanical explanations.[52] In fact, Einstein would later show by his theory of relativity that the laws of Newtonian mechanics had to be transformed in order to reconcile them with the laws of electricity and magnetism. At least during the first half of the nineteenth century, electricity was more closely associated with the newly emerging Romantic philosophy than with mechanical philosophy.

Romanticism and Naturphilosophie

In the philosophy that arose from the Romantic movement, the organic ideas associated with vitalism began to be applied to physical as well as biological phenomena. Although mechanical philosophy came to dominate eighteenth-century thought, especially in France, Great Britain, and Holland, interest in an organic or vital world view continued to be strong in the Germanic lands. What came to be known in the nineteenth century as *naturphilosophie* (nature philosophy) had its roots in the earlier philosophical ideas of G. W. Leibniz and Immanuel Kant.[53] Although Leibniz was associated with mechanical philosophy, he was particularly influenced by iatrochemical theories because of his close friendship with van Helmont's son. Like the mechanical philosophers, he believed that the world was composed of a number of discrete units, what he called monads, but unlike those philosophers, he did not distinguish between extended bodies and the world of spirit (or mind). The monads were centers of vital force that combined the passivity of matter with the activity of mind. Unlike mechanical atoms that underwent change because of external causes, a world composed of monads underwent change through its own inner growth.

The most direct influence on *naturphilosophie* was the work of Immanuel Kant. His Copernican revolution in philosophy established the framework for the Romantic world view in the nineteenth century. Kant, unlike earlier philosophers, did not believe that order was an inherent property of the external world. Rather he argued that the human mind actively ordered the experiences it received from the world. Kant's "revolution" was an attempt to overcome the mechanical duality between matter and spirit. In his *Critique of Pure Reason* (1781), Kant argued that the concept of substance was simply the end of a series of perceptions that began with intuitions of the mind. The basis of our perception of the external world (and therefore our only knowledge of it) is "through the forces which work . . . either by drawing others to it (attraction) or by preventing penetration (repulsion)."[54] In overcoming the duality of spirit and matter, Kant had essentially destroyed the mechanical concept of the material world and replaced it with a world based on attractive and repulsive forces. As he argued in his later work on the *Metaphysical Foundations of Natural Science* (1786), attractive or repulsive forces filled the universe; all phenomena in the natural world could be explained as a conflict between these types of forces. That is, at every point in space there were some attractive and repulsive forces whose interactions or conflicts produced the phenomena of the material world. Some conflicting forces would manifest themselves as material bodies while others might appear as gravitation, electricity, or magnetism. By eliminating the concept of matter and reducing phenomena to the underlying unity of forces, Kant removed the problem of the duality of spirit and matter and opened the door to a characterization of the world in terms of nonmaterial or spiritual concepts.

The *naturphilosophen* believed that all processes in nature took place through a polar (attractive and repulsive) interaction between mind and matter. These philosophers rejected the mechanical–atomistic explanations of the Newtonians and sought to replace them with dynamic and organic concepts that would transcend the mechanistic distinction between mind and matter.

One of the most influential of the *naturphilosophen* was Friedrich Wilhelm Joseph von Schelling. Although Kant had placed a new emphasis on the mind in ordering our experiences of the world, he still retained a distinction between subject and object. But many of his followers believed that his Copernican revolution had to be carried through to its conclusion. If Kant's philosophy were to be consistent, the objects in the external world would have to be regarded as products of the mind. This is not to say that the physical world can be interpreted as a product of the conscious activity of the human mind or that it is even a product of the unconscious mind. In fact, the source of the objective world was to be found not even in the individual mind at all. Rather it was to be found in a supraindividual mind, an Absolute subject. Therefore, the individual mind or subject and the external world of nature must be intimately connected to each other through the existence of an Absolute mind. Just as there can be no subject without an object and no object without a subject, the mind and the natural world are two components of an essentially unitary concept.[55]

For Schelling, nature was the objective manifestation of the Absolute and at the same time an expression of the mind or spirit. The role of the philosopher and scientist is to show how matter begins as force and is carried through dynamic processes finally to become mind.[56] Schelling believed that the natural world must be a purposeful system where the goal and function of the lower levels are to be found in the higher levels. All parts of nature, even those that would be classified by others as inorganic, such as inert matter, are seen by Schelling to be organic since their ultimate goal is to develop into life or spirit. As the philosopher Frederick Copleston notes, "It is thus truer to say that the inorganic is the organic *minus* than that the organic is the inorganic *plus*."[57]

For Schelling the world existed as a result of a dynamic interaction between subject and object, and the fundamental processes of nature must therefore be caused by the tension arising from polar forces. At the lowest level, matter itself consisted of an interaction between attractive and repulsive forces. Our experience of matter as something that occupies a space is actually the experience of a region of space filled with repulsive forces, but these repulsive forces could not exist within a defined space without the balancing effect of attractive forces. At higher levels of organization, these same forces could manifest themselves as other phenomena such as electricity and magnetism, chemical reactions, and finally as organic life.

This is not to say that *naturphilosophen* like Schelling were reducing organic life to a set of simple physical forces. All natural phenomena, including those we would classify as inorganic, were seen by Schelling and his followers as concrete manifestations of the Absolute, and therefore "alive." Essentially, the *naturphilosophen* were attempting to create a science based on vital forces rather than on the mechanical forces of Newton.

Although *naturphilosophie* did not become the dominant philosophy of science, it had a significant effect on nineteenth-century science.[58] For example, the scientific work of the German physicist Johann Wilhelm Ritter was strongly influenced by the Romantic philosophy of *naturphilosophie*. Ritter was educated at the University of Jena, a center of Romanticism, where he became friends with writers such as Novalis. After William Herschel discovered infrared light in 1800, Ritter discovered an analogous set of invisible rays at the higher end of the spectrum. He was led to this discovery of ultraviolet light by a tenet of Romantic philosophy, namely, that of polar oppositions in nature. If a set of invisible rays were found at the lower end of the spectrum, it was natural to look for their polar counterparts at the higher end of the spectrum. For Ritter, the most important aspect of both sets of invisible rays was that they could bring about chemical activity. When silver chloride was exposed to ultraviolet light it began to decompose. When the blackened silver chloride was exposed to infrared rays, however, the decomposition was reversed.[59] Such discoveries supported the idea that there was an underlying unity to the phenomena of light, heat, electricity, and magnetism, and that these various phenomena might be convertible into one another.

The actual relationship between electricity and magnetism was discovered by the Danish physicist Hans Christian Oersted. While at the University of Copenhagen, he was introduced to the ideas of *naturphilosophie* through a study of Kant. During a visit to Germany, Oersted worked with Ritter, attended the lectures of the philosopher Johann Fichte, and studied the writings of Schelling.[60] For twenty years he searched for some experimental proof of the underlying unity of electricity and magnetism, and in 1820 finally found a connection. During a lecture-demonstration Oersted passed electricity through a wire and noticed that a nearby compass needle was deflected. The discovery that an electric current could create a magnetic field caused a great sensation in the scientific world.[61]

The ideas of *naturphilosophie* were part of the basis of a new view of physical phenomena founded not on materialistic and mechanical properties but on the concepts of energy and fields. Although he could not be called a *naturphilosopher*, the British physicist, Michael Faraday, was influenced by Kantian and Romantic ideas during his youth.[62] L. Pearce Williams, a historian of science, has argued that these ideas provided the philosophical framework for Faraday's discovery of electromagnetic induction (the discovery that a changing magnetic field can induce an electric current in a wire), and his theory that electrical, magnetic, and gravitational interactions between bodies were the result of some *field* of force.[63] Faraday's concept of a field rejected the mechanical notion that matter was limited to a geometrically defined space, and argued instead that matter, through its association with fields, was omnipresent throughout space.[64]

Naturphilosophie attempted to unify the sciences, both physical and biological, by providing an organic basis for all of the sciences. One of the best examples of this approach was the work of Lorenz Oken, one of the leading *naturphilosophen* and editor of the journal *Isis*.[65] In his *Lehrbuch der Naturphilosophie* (1810), which had a significant impact on American transcendentalists like Ralph Waldo Emerson, Oken tried to show that mechanical and physical phenomena were lower level manifestations of a pervasive and vital World Spirit that was most obvious in organic nature. For Oken, the "great chain of being" did not end at the lowest form of life but continued into the physical and mineral world, comprising chemical substances, which had some self-activity as shown by spontaneous chemical processes, and finally mechanical matter, which had the lowest level of self-activity and individuality.[66]

For Oken, the physical and biological worlds came together through the phenomena of galvanism. A galvanic pile of zinc and silver was not an organism since the galvanic process was taking place only at a specific place, but the pile had a level of vital activity since it required air just like a living creature. If the galvanic process took place at every point in some material, that body would have a higher degree of self-activity and would become alive. He noted that an "organism is galvanism residing in a thoroughly homogeneous mass. A galvanic pile, pounded into atoms, must become alive."[67]

For the *naturphilosophen* like Oken, the phenomena of the world, both biological and physical, were united through their common connection to an all pervasive World Spirit. The higher activities of life could not be reduced to the lower level of mechanical activity. Rather, the World Spirit manifested different aspects of its own inner development which were externalized as different levels of nature. The individual grades of nature had no physical or historical connection to each other, and therefore could not be reduced to each other. But they all shared a common unity through their connection to the World Spirit. Here an understanding of phenomena was not accomplished through reductive analysis, but by means of taxonomy. Diversity was preserved but an underlying order was discerned when the diverse objects and phenomena were ranked according to a classificatory system.

Theories of Classification

The use of taxonomy, in place of reductive analysis, as a method of understanding was an important aspect of the organic world view.[68] Although classification systems serve the practical purpose of identifying and indexing information, they also serve the more ideal function of bringing order to some diverse set of objects.[69] Since at least the time of Aristotle, there has been an interest in classifying the natural world, but before the Renaissance "voyages of discovery" only a limited number of plants and animals were known to European naturalists. The discoveries of new plants and animals from newly explored regions of the world brought an increased interest in taxonomy during the eighteenth and early nineteenth centuries.

Most of the early classification systems were inconsistent and emphasized the identification of individual properties rather than establishing relationships between species. Such systems were artificial since they assumed that species could be placed into discontinuous and distinct categories based on one or two external characteristics. The most famous artificial system was the binomial scheme put forward by the Swedish botanist Carl Linnaeus, in which the number of stamens determined a plant's class and the number of pistils its order.[70] Linnaeus's system popularized the observing and classifying of nature. Even amateur naturalists could learn the binomial system and could enter in the widespread search for new species.

But Linnaeus himself was aware of the artificiality of his system and throughout his later life he attempted to discover a natural method that would express the inherent relationships between species. By the late eighteenth century, artificial systems were being challenged. Especially in France, naturalists, such as Georges Buffon, Michel Adanson, Antoinel Laurent de Jussieu, Charles Bonnet and Jean Baptiste Lamarck, believed that by use of multiple characteristics, species could be assigned to natural families.[71]

In his book *The Order of Things,* the French historian Michel Foucault suggests that this change in classification systems reflected a change in thinking about the world.[72] He argues that the relationship between visible structure and the criteria of identity was modified at the turn of the nineteenth century: "Throughout the eighteenth century, classifiers had been establishing character by comparing visible structures, that is, by correlating elements that were homogeneous."[73] But by the beginning of the nineteenth century, classification moved away from the visible and toward an internal principle that would relate the visible characteristics to a deeper cause. This new natural system of classification was based on *organic structure*.[74]

The most important element of a taxonomy based on organic structure was that basic characteristics of the plant or animal had to be linked to some essential *functions* which then became the basis for classification. According to Foucault, organic structure "subordinates characters one to another; it links them to function; it arranges them in accordance with an architecture that is internal as well as external, and no less invisible than visible."[75] In the older mechanical artificial systems, an object was classified by its most obvious feature, something independent of other characteristics, but in an organic system the superficial characteristics would have to be linked to some possibly hidden characteristic that performed the essential function of the object. This system of classification imposed a new order on nature. As Foucault has noted, "Animal species differ at their peripheries, and resemble each other at their centres; they are connected by the inaccessible, and separated by the apparent."[76]

Evolution

The interest in classification inevitably led to speculation concerning the evolution of species. Although Christian doctrine based on the Creation story in Genesis supported the idea of a fixed and unchanging world, the classification of geological strata and the increasing discoveries of numerous fossils of extinct species questioned the static world view of the eighteenth century.[77] By the mid-eighteenth century, several French natural scientists were already challenging the idea of an unchanging world. Jean Baptiste Robinet speculated that the great chain of being was not static but represented a hierarchy through which species had evolved because of some internal vital force. Another French naturalist Charles Bonnet argued that the changes in species resulted from some worldwide catastrophe, such as the Mosaic flood, which destroyed living creatures but allowed the germs of life to survive and regenerate as higher forms of life when the catastrophe ended. In England, Erasmus Darwin, grandfather of Charles, popularized the notion that the natural world exhibited development and progress.

Almost everyone associates the theory of evolution with Charles Darwin, but there were several evolutionary theories before Darwin's.[78] These earlier theories contributed significantly to an organic world view and had a continuing influence even after the publication of Darwin's *On the Origin of Species* (1859).[79] The most significant pre-Darwinian theory was advanced by the French naturalist Lamarck in his *Philosophie zoologique* (1809).[80] Believing in a separate "chain of being" for plants and animals, he held that this chain provided a path by which an organism, spontaneously generated out of inorganic material, could progress from the simple to the complex. An organism would move up the chain of being for two reasons. First, Lamarck believed that an "inner life force" drove an organism toward some higher level of perfection. If this inner force were the only reason for evolution, it would result in a single line of development, however, Lamarck's observations of the natural world made him realize that organisms deviated from a single line of development.

To explain such deviations, he proposed a second cause for evolution. Lamarck argued that the changing environment would result in certain "needs" (*besoins*) that would lead to new habits on the part of the organism. These habits would cause nonmaterial or imponderable fluids, such as electricity, to bring about changes in bodily organs, which could then be inherited by the offspring. In his well-known example, the giraffe's long neck evolved because of the animal's continued stretching to reach the higher leaves on the tree. This striving on the part of the animal was more unconscious than conscious, the result of a *sentiment intérieur* or inner sense. Through the "inner force" and the "inheritance of acquired characteristics," Lamarck was able to transform the static great chain of being into an escalator on which organisms continually emerged from inorganic material and evolved toward some "perfect organization." Lamarck's theory of evolution reflected an important characteristic of the organic world view—the idea of teleology. Like many other naturalists and physiologists, he saw the world as being directed by some end or goal, toward which the world was moving.

The idea of purposeful development was an important part of organic thinking and even theories that challenged teleology were sometimes interpreted in a teleological way. The most explicit challenge to a teleological view of nature was articulated by Charles Darwin.[81] With his *On the Origin of Species,* Darwin made two contributions to a theory of evolution. First, using data he collected as a naturalist on his journey around the world on the *H.M.S. Beagle,* he was able to make a convincing argument that evolution was, in fact, a reasonable and scientifically supportable theory to explain the temporal and geographical distribution of plants and animals. This aspect of Darwin's work has never been seriously challenged by any other scientific theory. But Darwin also made another more controversial contribution by putting forward a specific nonteleological mechanism to explain how evolution occurred. He rejected Lamarck's inner force and instead made the role of the environment the sole-determining factor. Darwin had noted that plants and animals produced more offspring than the world could sustain and that certain random variations in these offspring could be inherited by future generations. Therefore, he postulated a more mechanistic theory of "natural selection," by which plants or animals with favorable variations would be "selected out" through competition for a limited food supply and would pass their variation on to their offspring.

Darwin's theory of natural selection had eliminated any reference to an inner-directing force and, consonantly, it also eliminated any dependence on a purposeful plan in nature. The divine plan of God unfolding in nature was replaced by the materialistic idea of "survival of the fittest" (a phrase coined by Herbert Spencer before Darwin's theory was published).

Many of the supporters of the idea of evolution found Darwin's mechanism of natural selection unacceptable.[82] Since a theory of genetics was unknown at the time, Darwin was never able to fully explain the source of the variations in offspring, leading a significant number of his followers, especially in France, Germany, and America, to speculate that these variations could have originated in some inner vital force or through the inheritance of acquired characteristics.[83] Thus although Darwin's theory of natural selection appeared to do away with vitalistic and teleological explanations, a significant number of Darwinians reinterpreted Darwin's theory in teleological or neo-Lamarckian terms. There are even hints that Darwin himself eventually came to accept certain neo-Lamarckian beliefs.[84]

The concept of evolution provided a unifying element to an organic world view by furnishing a framework in which embryology, taxonomy, morphology, paleontology, physiology, and geographical distribution were shown to be interrelated. But the elevation of evolution to a universal principle, applicable to all phenomena, resulted in a modification and reinterpretation of the organic world view.

Organic Theories of Society

During the nineteenth century evolutionary ideas led to a new model of the relationship between organic theories and mechanical explanations of the world. One of the best examples of this new understanding was in the area of social philosophy. The success of evolutionary theories in the natural sciences led to the belief that evolution or development was a universal principle that could equally be applied to social phenomena.[85]

One of the most influential thinkers to apply organic and developmental ideas to society was Auguste Comte, who laid the groundwork for the modern discipline of sociology.[86] A fundamental aspect of his social philosophy was the "law of three stages" and the "hierarchy of the sciences." Comte believed that thought and society evolved through three stages of development. In the theological stage, explanantions of phenomena relied on some divine agency, while in the meta-physical stage, hypothetical entities were the basis of explanation. The final positivistic stage would reject all theological and metaphysical concepts and explain phenomena on the basis of "sense-reports," which would have the status of scientific "facts." But not all forms of human knowledge progressed through the three stages at the same time. So Comte proposed a hierarchy of knowledge, beginning with mathematics, which was the first area to become a truly positive science, and continuing through astronomy, physics, chemistry, biology, and finally sociology, a term coined by Comte to represent a positive science of society. Within the hierarchy, each science was dependent on all of the preceding sciences. That is, the science of sociology was dependent on the concepts and laws of biology, but in its turn, biology had to be seen as dependent on, although not reduced to, the concepts and laws of the inorganic sciences such as chemistry, physics, astronomy, and mathematics. For Comte, the laws governing society emerged through a kind of evolutionary development from the laws of biology, but those laws themselves emerged from more materialistic or mechanical laws.

Comte's evolutionary and organic model of society was shared by the British political philosopher Herbert Spencer.[87] Beginning as a railway engineer, Spencer became interested in the idea of evolution before the publication of Darwin's work, and he turned to philosophy, becoming one of the most popular writers of his time. Although Spencer coined the phrase "survival of the fittest" which Darwin then used in his own work, he continued to support a more Lamarckian theory of evolution. Throughout his life, Spencer labored to create a "synthetic philosophy" based on the application of evolution to all phenomena, including the physical, the biological, and the sociological. Spencer believed that human societies evolved under the same laws that governed biological evolution. Like organisms, societies moved from an earlier stage in which there was a great deal of homogeneity, to a later stage of social differentiation. The evolution of a simple society in which all individuals were engaged in similar activities into a complex society in which individuals take part in highly specialized activities, parallels the evolution of a simple one-celled organism into an organism with a multiplicity of highly specialized organs.

After the publication of Darwin's work, Spencer's ideas came to be labeled Social Darwinism and as such influenced the social philosophy of people like William Graham Sumner, who argued that competition and survival of the fittest were the basic processes by which societies progressed.[88] Because fitness could be interpreted as financial success, political power, military power, or social status, Social Darwinism had wide appeal among nineteenth-century businessmen and politicians, such as John D. Rockefeller, Andrew Carnegie, and Theodore Roosevelt, who used it to justify the establishment of powerful monopolies and imperialistic foreign policies.

For Spencer, evolution was not only a property of biological and social systems, but a universal principle that applied to inorganic phenomena as well. In *First Principles* (1860), he argued that the emergence of our solar system out of a large gaseous disk of matter and the development of the Earth's crust out of a molten sphere, were both examples of evolutionary processes working in the inorganic world.[89] According to Spencer's theory, organic and inorganic systems were related to one another through the law of evolution. Both systems moved from states of incoherent homogeneity to states of coherent heterogeneity.[90] But as one moved from inorganic systems to organic and even to social, or superorganic systems, one discovered an increasing degree of mutual dependence, or cooperation, between the component parts of the system.[91]

Comte's and Spencer's visions of an organic model for society became most fully realized in the work of Emile Durkheim, one of the founders of the discipline of sociology.[92] In *The Division of Labor* (1893), Durkheim argued that primitive societies were based on a kind of social cohesion that was essentially mechanical. The simplest forms of human societies functioned like a group of unconnected units or cells, having little dependence on each other; various parts of a society could be considered as self-sufficient units that were related to the whole in a mechanical way.[93] But as societies began to evolve, various segments differentiated into clans or kinship groups. Eventually societies evolved into a model based on what Durkheim labeled "organic solidarity," in which individuals or groups within a society were interdependent on one another. Differences among the individual units made specialization of function an important characteristic of this society.[94]

The work of Comte, Spencer, and Durkheim reflected a new way of understanding the organic world view and its relationship to the mechancial world view. The two views came to be seen not as opposed but as different evolutionary stages of human thought, with the organic view emerging out of the mechanical view. There were still significant differences between the two just as there were differences between humans and monkeys. The relationship did not imply that the organic could be reduced to the mechanical any more than a human can be reduced to a monkey, but the idea of evolution led to reinterpretations of both views and began to blur the line dividing them.

Conclusions

Like the mechanical world view, the organic world view developed and changed throughout its history. Several problems had led to modifications of the more dogmatic interpretations of organic philosophy and, by the nineteenth century, it was coming closer to the mechanical world view. The idea of a vital substance, spirit, principle, or force proved to be a continuing problem for organic philosophy. Many vitalists resorted to the position that the vital power which organized and activated living organisms was similar to the Newtonian forces which activated mechanical matter. Other vitalists came to associate the vital principle with electricity. In associating vitalism with gravity or electricity, the organic world view was incorporating concepts and explanations that could not easily be classified as either mechanical or organic since both gravity and electricity defied traditional mechanical explanations. The idea of evolution introduced other problems for organic philosophy. Darwinian evolution questioned the concept of teleology that always had been at the heart of the organic world view. Even in its teleological versions, evolution raised the possibility that organic systems were not opposed to mechanical systems but simply a later evolutionary stage.

Despite modification and change, one set of ideas characterized the organic world view in all of its forms.[95] In general, the organic world view assumed that the world functioned like an organism such as a plant. Phenomena were explained in terms of a purposeful design or plan that governed the organization of the system and interrelationships between its parts. These parts were not autonomous or independent entities but had meaning only in so far as they could relate to the whole. Change did not have to be explained as some special situation; growth, development, transformation, and evolution were inherent characteristics of all phenomena. In this world view an explanation of phenomena must in the end be a description of some purposeful behavior or goal-seeking activity.

The organic world view was applied not only to the biological world but to the rest of the physical world as well; organic concepts were extended beyond the biological words and used to explain technological processes. The symbol of the organism or plant enabled technology to be understood as what could be called organic machines.

5

Organic Machines: Technology as Plant

The human body is the magazine of inventions, the patent office, where are the models from which every hint is taken. All the tools and engines on the earth are only extensions of its limbs and senses.

RALPH WALDO EMERSON, *Works and Days*

The organic world view provided a different model for understanding machines and technological processes from that provided by the mechanical world view. The organic view emphasized the role of process and change rather than reductionism and analysis. Also unlike the mechanical world view, in which designs and rules were imposed from the outside, the organic world view postulated an internal design principle and a set of inherent rules that unfolded during the process of development. As we shall see, the organic view led to an entirely different concept of the machine and the design process. At first the organic world view served only as a model for technological processes. But as this world view became more accepted and widespread, supporters as well as critics of technology no longer saw organic life as simply a model for technology; rather the machine began to incorporate some form of life itself. While the machine might not appear to be alive, opinion held that technology was based on and designed to release the vital forces that were part of organic nature.

Organic theories of technology proliferated during the nineteenth century, and, as with Romantic philosophy, earlier thought influenced the development of an organic concept of technology. In many cases these ideas were the same as those connected with the mechanical concept of life in the seventeenth and eighteenth centuries. We have already seen how the creation of clocks and automata supported the attempt to explain organic life itself in terms of a mechanical philosophy. But, as will be shown, these same creations could be used in support of the idea that machines had a life of their own and that technology could be stimulated by attempts to imitate the essentially organic processes of human, animal, and plant physiology.

Organic Origins of Technology

The idea that technological processes are essentially organic has such a long history that it is impossible to locate its origin. It has been argued that the first tools were themselves organic: animal bones, teeth, and antlers.[1] There is ample evidence that by the Neolithic period many of these natural items were being used as clubs, knives, or cutting tools. Many early Stone Age tools, if not actual animal parts, can be seen as having been modeled on them. Knives imitated the cutting action of the front teeth, grinding stones imitated the back teeth, and clubs imitated and extended the action of the arms. Many early cooking utensils were also either taken directly from nature or designed as imitations of natural products. Cups, for example, and vessels for storing food and water were, if not actual gourds, then imitations of them.

The discovery of agriculture led to new connections between tools and the organic. Some scholars have argued that people's recognition of the importance of agriculture led to a new organic view of the world.[2] The processes of planting, growing, and harvesting provided analogies in nature for understanding such things as the phases of the moon, human fertility, and the change of the seasons. The natural analogies between agriculture and human reproduction caused many agricultural tools and implements to take on sexual connotations.

One of the clearest examples of the relationship between technology and the organic is found in the myths associated with the discovery of metals and the development of metallurgy. Many of the myths concerning the origin of metals were based on the belief that metals and minerals contained some living force and developed in the earth in a manner similar to the development of a fetus in the womb. This belief had several consequences. Metals that had to be mined were seen as not yet fully developed; their extraction from the earth cut short the developmental process. Consequently, the metallurgical procedures such as hardening, smelting, and alloying had to be seen as a continuation or acceleration of the natural organic process the metals would have gone through had they been left in the earth. Mircea Eliade quotes an eighteenth-century writer:

> What Nature did in the beginning we can do equally well by following Nature's processes. What perhaps Nature is still doing, assisted by the time of centuries, in her subterranean solitudes, we can make her accomplish in a single moment, by helping her and placing her in more congenial circumstances. As we make bread, so we will make metals.[3]

If the procedures of metallurgy were more akin to acts of procreation than fabrication, the tools used in these procedures must also take on an organic quality.[4] The primary tool of the metallurgist was the furnace. It was used to reduce ores into metals, to harden metals, and to smelt together two different metals. Since the basic purpose of the furnace was to accelerate and perfect the growth of the metallic ores extracted from the earth, it took on the character of an artificial womb in which the gestation of the ores could be completed. Most early furnaces were consciously designed to resemble either an egg or a uterus.

In many ancient myths and rituals there was an even closer connection be-tween metallurgical procedures and the organic. Defining metallurgy as an act of procreation, some myths argued that something containing life could develop only from another life form. One result was a belief that metallurgical processes had to be accompanied by some type of blood sacrifice. The sacrifice was required so that the organic force of the victim could be transferred to the metals themselves and provide them with the essential element they needed for further development and growth.

Mircea Eliade produces evidence that the ancient Babylonians conducted some type of sacrifice when a new furnace was built.[5] Although there is a controversy over the translation of an ancient text, some scholars argue that the Babylonians placed aborted fetuses into the furnace to assure its future success. Whether this was true or not, there are recent rituals in areas of Africa where aborted fetuses are sacrificed to the furnace. Parallels have been drawn between an aborted fetus and the processes of metallurgy, which bring about a birth of metal before its time.

Not only was there a connection between the furnace and the organic, but the tools associated with the metallurgist and the smith were thought by some to be alive.[6] In primitive and ancient societies, metallurgical tools, such as a hammer, anvil, and bellows, were believed to have magical powers and the ability to work by themselves. In some cultures the hammer of the smith is treated like a special child. There were also many sexual taboos and rituals associated with the making of metallurgical tools.[7] While they are being produced the smith must abstain from sexual activity; on completion he must have sexual relations so that the tools can be made to come alive.

This brief examination of the history of metallurgy, one of the oldest technolo-gies, reveals an intimate relationship with the organic. The rites, myths, and symbols associated with metallurgical procedures were embryological. Early metallurgists believed that if they were going to intervene in the processes of nature, intervention must be modeled on them. Many of these specific beliefs continued well into the eighteenth century, providing a model for a technology based on organic symbols rather than mechanical ones.

Aristotelian Theories of Technology

More generally, an organic theory of technology was developed by Aristotle. His philosophy, which established some of the key elements of the organic world view, also provided a framework for a technology based on organic processes. Much of Aristotle's philosophy was associated with the theoretical sciences such as mathematics and physics or the practical sciences such as ethics and politics, but it also included the productive sciences which in turn comprised not only topics such as rhetoric and poetics, but also technology and art.[8] For Aristotle, a theory of art was broad enough to consider painting, sculpture, poetry, drama, and technol-ogy. In classical Greece, the term *techne,* which is the root of our word tech-nology, applied to both the fine arts and the practical applied arts of tech-nology.

Aristotle believed that an object was brought into being when a form was manifested in some material substance. *Techne* meant the imposition of form onto material. The processes of *techne* or art were similar to another process of creation, that of nature. In both art and nature something is brought into being (*genesis*), but in nature the material itself generates its inherent form, like an oak coming from an acorn, while in art the new form is imposed onto some material by an external agent or artist. In nature design is internal, while in art design is external.

Although Aristotle made a distinction between the processes of nature and the processes of art, he saw an intimate connection between the two. Even though a tree is made out of a seed by nature,

> if a house were a thing generated by nature—if it grew—it would be produced in the same way in which art (*techne*) now produces it. If natural things were not produced by nature alone, but by art also, they would be produced by art in the same way that they are produced by nature.[9]

That is, both art and nature create objects by realizing the potential that is in the material being used. In nature the agent that causes that potential to be realized is also inherent in the material, while in art the agent is an external craftsman or builder.

The action of nature and the action of *techne* are not distinct and independent processes; rather they function in very compatible ways. Since nature, by herself, cannot grow a house out of wood and bricks, she must accomplish it through some outside agent such as a builder. According to Aristotle, ''In general, then, art (*techne*) in a sense completes what nature is unable to finish, and in a sense imitates nature.''[10] As the classical scholar John Herman Randall points out, Aristotle does not mean that *techne* can imitate the products of nature, such as a tree or a bird, but rather that art imitates the processes by which nature produces her products. Aristotle does not discuss in detail how *techne* should ''imitate'' the processes of nature, but Randall argues that a few things are clear.[11] The process of imitation reflects Aristotle's belief that an object is ''brought into being'' when some form, for example a sphere, is manifested in some material, like bronze. In the process of imitation, the craftsman takes a form that exists in nature, abstracts it from its natural material, and realizes it in some new material of his art. This does not mean that the craftsman simply copies the shapes and designs of nature. The craftsman's form is ideal and abstract; the new material he is using will impose new conditions on the form that were not there in nature. For example, a builder might get the idea of creating a shelter by observing a natural stone cave, but when'he tries to create the form of a cave in a different material, wood, that material will have certain structural requirements that will make a wooden house look very different from the original cave.

Aristotle provided a model for an organic approach to technology, one in which nature provides the forms that technology is to *imitate,* and technology helps to *complete* the processes that nature by herself could not accomplish. Although not all ancient Greeks saw technology working in harmony with nature, the historian R. Hooykaas has pointed out,

This idea that the arts took their origin from an imitation of Nature was a popular theme with the poets and philosophers of antiquity. Democritus said that the spider taught us spinning and weaving, and the swallow building. Lucretius thought that the art of cooking was inspired by the sun; while, according to Vitruvius, observation of the rotating heavens led to the making of machines. [12]

Organic Theories of Technology in the Middle Ages

From the late Classical period and through the Middle Ages and Renaissance, there was great interest in the attempt to use technology to imitate the processes of nature, especially the self-movement of birds, animals, and humans. In Chapter three we discussed the attempts to create automata by people such as Hero of Alexandria and others. Many of these automata were used later by the mechanical philosophers to support the view that organic life was essentially mechanical. But, as we shall see, it is also true that such attempts could be used to support the idea that technology is essentially organic.

Although the automata developed during the Middle Ages and Renaissance appear to be mechanical, we must be very careful in classifying them as such without investigating what it meant to be mechanical. As pointed out earlier, the mechanical world view did not become fully formulated until the seventeenth century. Although there were earlier inventions such as clocks and automata, which helped support the mechanical world view, the medieval and Renaissance view of mechanics included elements that could also be seen as organic.

In the Middle Ages there were attempts to show that the practical arts, such as mechanics, could help mankind move from an incomplete state toward some state of divine completion in a way similar to the traditional role of the seven liberal arts (grammar, rhetoric, dialectic, geometry, arithmetic, astronomy, and music). [13] One attempt came from the twelfth-century scholar Hugh of St. Victor, who argued that the mechanical arts must be imitations of nature. [14] He gave two examples:

> He who builds a house bears in mind the conformation of a mountain. . . . The summit of a mountain offers no resting place for waters; similarly a house must be built to a certain height in order to afford security against the impact of invading rain. The discoverer of the use of clothing had previously observed that every form of life has its own method of protection to ward off damage from Nature. Bark surrounds the tree; feathers cover the bird; fish are enwrapped by scales; sheep by wool. [15]

Although plants and animals were provided protection through the designs built into nature by God, Hugh argued that humans' lack of protection allowed them to use their reason to complete themselves through the observation of nature. By associating the mechanical arts with human beings' movement toward divine completion, Hugh of St. Victor and other medieval scholars subsumed these arts within the overall divine plan of nature. Because the mechanical arts were based on an imitation of nature, they were connected to the organic. Therefore, even if an object such as an automaton was understood to be a mechanical creation, this did not necessarily mean that it was seen as distinct from or opposed to organic life.

Although Hugh of St. Victor appreciated the connection of technology with organic nature, he, like many other medieval writers, concluded that technology was limited to copying natural models and could never duplicate nature's product.[16] But as William Newman has recently argued, many medieval alchemists believed that their techniques, although imitations of nature, could actually equal and even improve on nature's own products.[17] He notes that such influential figures as Roger Bacon argued that metals could be transmuted into other metals by alchemical techniques.

Technology and Renaissance Magic

Another element linking the mechanical to the organic was belief in magic. Beginning in the Middle Ages, but reaching a peak during the Renaissance, there was an effort to associate technology with magical powers.[18] Since these magical powers were derived from the secret forces in nature, technology also became linked to the organic. In 1269 Pierre de Maricourt wrote one of the first studies on the magnet, which was thought by many at this time to be associated with magical powers and even to be alive since it would move to orient itself in a north-south direction. Maricourt argued for a close connection between magic and technology: "For in occult matters, we investigate many things by manual industry, and in general without it we are unable to bring anything to completion."[19] That is, technology was the means by which magical or occult ideas were made to yield actual results.

One of the most common uses of technology to bring about or release the magical powers inherent in nature was the attempt to reproduce life itself by means of technology. We have seen how clockworks and automata contributed to a mechanical world view; the automata of the seventeenth and eighteenth centuries were rooted in the mechanical philosophy of Descartes, which emphasized a distinction between spirit and matter. Because the technology behind these automata was seen as purely mechanical, they symbolized a view of organic life that was also mechanical. But the magical automata of the Middle Ages and Renaissance supported a very different point of view. Although the technology that went into their making was mechanical, it was an external manifestation of a deeper magical or spiritual force that was at the root of all life. Therefore, the technology of these automata was closely associated with magical, spiritual, or organic forces.

A prime example of an automaton based on an organic world view is the golem. Although the golem was not created by technology, the myth of the golem had a great influence on the development of a magical-organic form of technology. The term *golem*, a Hebrew word meaning an unformed amorphous substance, goes back at least as far as biblical times and was in fact used to refer to Adam.[20] But the myth of the golem was formulated during the Middle Ages and Renaissance. In the Jewish legend, a golem was created by a learned rabbi molding earth or clay into the shape of a human being. The figure was then brought to life by the performance of an esoteric ritual, usually involving the application of the names of God and the Hebrew letters that were originally used in the creation of the universe. In most versions of the legend, the magical ritual simply releases the animistic forces that are already inherent in the material of the world. Although the golem begins as a simple mute servant, the creator almost always loses control and must try to return the golem to its inert state. By the sixteenth century, some versions of the legend presented the golem as a mechanical figure of wood and hinges.[21]

By the Renaissance, the legend of the golem combined with the creation of mechanical automata in the development of a technology that was closely associated with spiritual-organic forces. Although there was a great interest in mechanical automata during the Renaissance, there was no distinction between spirit and matter that would exist after Descartes. As historian Frances Yates has noted, there was a basic confusion "between mechanics as magic and magic as mechanics."[22] This attitude is best portrayed by Henry Cornelius Agrippa in his *De occulta philosophie* in which he outlined three levels of magic that were common during the Renaissance.[23] The lowest level was natural magic, and it operated in the natural world through the use of sympathies and antipathies, such as the use of smoke to make rain. The highest level of magic was religious magic, which functioned in the supercelestial world through rituals and resulted in the conjuring of angels and other spirits. These two levels are what we commonly characterize as magic today. But during the Renaissance there was a third level of magic that existed between natural magic and religious magic. This level of magic, labeled mathematical magic, operated in the celestial world and through the use of mathematics, number mysticism, kabbalistic practices, and mechanical technology. For many people in the Renaissance, mechanics and technology were a form of mathematical magic that functioned in a way similar to the sympathies and antipathies of the natural world and the spirit conjuring of the supercelestial world.

This magical view of mechanics and technology was supported by the rediscovery during the fifteenth century of a set of writings attributed to an ancient Egyptian soothsayer, Hermes Trismegistus. Although it was later learned that the works were written during the first or second centuries A.D., they had great authority during the Renaissance. They outlined a magical or Hermetic philosophy based on astrology, alchemy, the Kabbalah, and mechanics. It was common to associate their story of magical statues, brought to life by the Egyptians, with the mechanical and pneumatic devices of Hero of Alexandria.[24] As Yates observes, this was not done to disparage the role of magic by explaining the Egyptian statues as simply technological. Rather, it reflected a belief either that the use of technology drew souls into the material of the world or that the invention of a mechanism released the world spirit (*anima mundi*) that was inherent in all material.[25]

This view of technology seems to have had some impact on one of the most famous figures of the Renaissance—Leonardo da Vinci. Although he is usually portrayed as a precursor of the Modern Age, there are indications that he was also in the tradition of the Renaissance magus, referring in several of his writings to "Hermes the Philosopher."[26] Leonardo's concept of force was closer to the spiritual ideal than to the mechanical. In a *Hymn to Force,* he says, "Force is nothing but a spiritual power, an invisible energy which is created and communicated, through violence from without, by animated bodies to inanimated bodies, giving to these the similarity of life, and this life works in a marvelous way, compelling all created things from their places, and change in their shapes.''[27] That is, for Leonardo, using technology to control the action of forces within machines or structures was analogous to giving life to an inanimate body. As technology was associated with organic life, so the engineer should function like a physician. In an application for a commission on the Milan Cathedral, Leonardo wrote:

> You know that when medicines are rightly used they restore health to the invalid, and that he who knows them well makes the right use of them if he also understands what man is, and what life is, and the constitution of the body, and what health is.
> . . . The case of the invalid cathedral is similar. It also requires a doctor architect who understands the edifice well, and knows the rules of good building from their origin.[28]

In the Renaissance the connection between technology and organic life also had roots in metallurgy. The processes of mining and smelting metals were closely connected to organic processes through the theories of alchemy, which postulated that metals grew in the earth like other organic substances. Many of the people associated with metallurgy and alchemy had training as physicians, which led them to view technology in terms of the medical and biological sciences. For example, the great physician and iatrochemist, Paracelsus, spent much of his early life wandering through mining areas where he learned much of his alchemy. He viewed every creative process, including technology, as based on the organic transformations of nature. Technology, like alchemy, simply brought about the completion of what already existed in nature. According to Paracelsus, "This completion [of nature] is called alchemy. For an alchemist is the baker while he bakes bread; the vintner while he makes wine, a weaver while he makes cloth. Therefore he who brings that which grows by Nature for the use of man into the state ordained by Nature, is an alchemist.''[29] The technology of Paracelsus was part of organic nature since it acted in harmony with the processes of nature.

The association of technology with the processes of nature (sometimes hidden) was a persistent belief in the Renaissance and had roots in the study of natural magic. One of the most influential books on the subject was *Natural Magick* by John Baptista della Porta, first published in 1558 and followed by numerous later editions in English, Italian, French and Dutch.[30] This book encompassed a wide range of subjects, including what we would call today magical ideas such as astrology and alchemy; old wives' tales such as the generation of animals from putrefied materials; folk remedies; household hints, for instance, on baking and preserving foods; scientific and technical subjects such as distillations, optics, and tempering steel; and finally women's makeup and perfumes. All of these, both the practical and the mysterious, were classified by della Porta as forms of natural magic. According to della Porta, natural magic was simply the "dutiful handmaid" of nature, and "if she finds any want in the affinity of Nature, that is not strong enough, she doth supply such defects . . . as in Husbandry, it is Nature that brings forth corn and herbs, but it is Art that prepares and makes way for them."[31] This definition of magic includes not only the traditional forms of magic such as alchemy and astrology, but also technology, so long as technology is seen as based on organic principles. Instruction in natural magic not only included philosophy, astrology, and chemistry but also put particular emphasis on the organic. Della Porta argued that the magician "must be a skillful Physician. . . . Moreover, it is required of him, that he be an Herbalist, . . . as there is no greater inconvenience to any Artificer, than not to know his tools that he must work with: so the knowledge of plants is so necessary to this profession, that indeed it is all in all."[32] Finally, the natural magician had to have the practical skill that we would associate with technology:

> He must be a skillful workman, both by natural gifts, and also by the practice of his own hands: for knowledge without practice and workmanship, and practice without knowledge, are nothing worth; these are so linked together, that the one without the other is but vain, and to no purpose.[33]

The Renaissance tradition of natural magic thus viewed technology as one of the arts that could be used to release or complete hidden potentials that existed in the world of nature.

Technology, Magic, and the Theater

One of the best examples of a magical view of technology is associated with the Elizabethan and Jacobean theaters. In *Theatre of the World,* Frances Yates outlines the close connection between the magical-Hermetic world view and the technologies involved in stagecraft.[34] A significant figure in this movement was John Dee, who was both an important scientist and the model of the Elizabethan magus. He was born in 1527 and, after being educated at Cambridge, became closely associated with Queen Elizabeth and her court. From a traditional point of view Dee has always been an enigma. At times he appears to have been a rational scientist, arguing for the use of mathematics and mechanics to advance science, while at other times he seems to be the model of Shakespeare's Prospero, with his interest in conjuring spirits and angels. But as Yates has argued, we have interpreted his science from a modern point of view. For Dee, mathematics and mechanics were part of the secret arts that revealed the spiritual nature of the world.

Dee first earned the label conjuror when he was a student at Trinity College, Cambridge. For a play by Aristophanes, he created a giant mechanical scarab which flew to the roof of the hall carrying a man on its back.[35] As Yates notes, it was in the theater, through the use of machines, pulleys, and trick perspectives, that technology and magic came together. Through the experiences of theatrical productions, the audience began to associate technology with wonderous and magical effects. But this was not magic in the modern-day sense of a trick or an illusion. Renaissance magic was the release of the real spirits that pervaded the world. This can be seen in one of Dee's most influential works, a preface to Henry Billingsley's English translation of Euclid, which was published in 1570 and became one of the most important scientific books of the Elizabethan age. In his preface, Dee outlined the uses of mathematics and mechanics, and he encouraged their improvements. Dee's preface, a manifesto for the advancement of science, was influential well into the seventeenth century.[36] Although sections of the preface seem quite modern with their treatment of geometry, hydrography, geography, astronomy, architecture, and navigation, these topics are treated along side more magical topics such as astrology and something Dee called "thaumaturgike."[37]

It is in Dee's discussion of thaumaturgike that we find an approach to technology based on magical principles. He defined the term as "that Art Mathematical, which giveth certain order to make strange works, of the sense to be perceived, and of men greatly to be wondered at."[38] The "wonder-work," as Dee explains, is brought about by technology such as the pneumatics of Hero of Alexandria, weights, strings, springs or by "other means" such as those used by Albertus Magnus to make a brazen head seem to speak.[39] As examples of what could be done by thaumaturgike, Dee noted the revolving model of the planets created by Archimedes; a wooden dove that could fly, made by Architas; a mechanical fly that flew about guests at a banquet in Nuremberg; and a mechanical eagle that was sent to greet the emperor at Nuremberg.[40] Dee thus placed the use of mechanics to make an object come to life in the same category as brazen heads that spoke or Egyptian statues that magically came alive. Like magic, technology brought things to life because hidden in all things, including what we call inanimate matter, was a world spirit (*anima mundi*). Therefore, this magical technology was, at its root, essentially organic even though it relied on mechanical devices.

Technology, Magic, and Architecture

At the basis of much of the Renaissance interest in an organic or magical technology was the revival of interest in the works and theories of Vitruvius.[41] From Leonardo da Vinci through John Dee, there was an interest in the relationship between art and science based on number and proportion. According to Vitruvius's *De architectura*, the divine proportions that underlay the design of temples and the universe itself were based on the human body. The Vitruvian figure of a man's body with arms and legs fully extended and placed inside a square and a circle became a common symbol in Renaissance art and architecture. This symbol was intended to show that the fundamental proportions of the circle and the square, which were used frequently in Renaissance architecture, were ultimately reflections of the proportions of the human body. The Vitruvian symbol shows that a fascination with geometric forms, such as we see in Renaissance art and architecture, need not exclude an organic or biological world view.

The person who carried the traditions of Vitruvius into the Jacobean Age was Robert Fludd, a physician who was a follower of John Dee and the iatrochemistry of Paracelsus.[42] Fludd, like John Dee, made a connection between technology, magic, and the theater, but he carried the relationship even further. As Frances Yates has shown, Fludd established a connection between theaters and the art of memory.[43] Since the time of antiquity, it had been common to use an actual building as a memory device. Images of ideas to be remembered were associated with parts or rooms of a building. Then the user or orator mentally went through the building in order to recall the ideas to be remembered. During the Renaissance, the art of memory became associated with a magical philosophy. Based on the analogy between the microcosm (human beings) and the macrocosm (the universe), Fludd's art of memory argued that the mental and spiritual ideas that existed in the mind and were organized by the memory were actually reflected in the organization of the external world.[44] Fludd specifically associated one type of memory system with the structure of theaters. Through a belief in the art of memory, many Renaissance technological structures, especially theaters, were seen not as creations out of some alien mechanical world, but as reflections of the inner world of the mind. Therefore, the Shakespearean statement that "all the world's a stage" was not a description of a world dominated by technology, but a statement about the mind's ability to actively reflect its order in both technology and the universe.

Another person who brought together interests in architecture, theaters, and technology was the great designer Inigo Jones, who was labeled the British Vitruvius. Jones was known for his gardens and buildings throughout England, including the Queen's House at Greenwich and the new facade of St. Paul's Cathedral, and was equally famous for his stagings of Shakespearean plays for the royalty of Europe. In these plays or masques, emphasis was placed on the ability of machines to create elaborate illusions and changes of scenes. In some productions, like Shakespeare's *The Tempest*, the magic of the production paralleled the magical content of the play. By using the magical technology of the theater, a character like Prospero could bring to life an airy spirit before the audience's eyes. As Frances Yates has stated, "In Prospero we may now see, not only the Magus as philosopher and as the all-powerful magician ushering in the scientific age about to dawn, but also the Magus as creator of the theatre and its magic."[45]

A center for the magical uses of technology was in Heidelberg, at the court of Princess Elizabeth, daughter of King James I of England and her husband Frederick V, elector of Palatine.[46] At Heidelberg magic and technology became closely associated with garden design. Salomon de Caus, a French engineer and friend of Inigo Jones and Elizabeth's brother Prince Henry, became architect and engineer in charge of improving the castle and its grounds. De Caus filled his gardens with water-powered mythological statues and elaborate fountains. Many of the statues made sounds, and music was said to be provided by an elaborate water-powered organ similar to one described by Vitruvius. The works of de Caus combined several important elements. First, his mechanical statues had magical associations. The ability to make a statue talk and move, even by mechanical means, was seen by some as a form of magic. Second, his works were closely associated with the theatrical technology of the Elizabethan stage that Frederick had seen in London and no doubt wanted recreated in his gardens at Heidelberg. Finally, and most importantly, de Caus's work established a close link between technology and the organic. Many, if not most, of his spectacular mechanical works were designed for use in gardens. In such a natural setting, their ability to move would be associated not only with magic but with organic life. The statues and fountains were not intended to be seen in opposition to the surrounding natural plants, flowers, and animals; rather the technological objects were to be seen as part of the organic world. During the sixteenth and seventeenth centuries the popularity of technological water-gardens spread throughout much of Europe, carrying with them this close association between technology and the organic.[47]

Organicism and the Steam Engine

The seventeenth-century interest in self-moving technological devices reached its height at the beginning of the eighteenth century with the development of the steam engine.[48] Several inventors, including Leonardo, della Porta, and de Caus, had described devices powered by steam. By the middle of the seventeenth century, several people were working on devices that used the vacuum formed by condensing steam in a closed vessel to raise water out of mines by suction. During the next fifty years this suction device or "Miner's Friend" developed into a new source of power. In 1661 the German engineer Otto von Guericke had shown that a vacuum inside of a cylinder fitted with a piston caused the atmosphere to exert a large force on the piston. At the end of the century, the French physicist and physician Denis Papin had demonstrated that a vacuum could be created inside a cylinder by condensing steam. Papin also invented a device called a "digester" which functioned like a pressure cooker. By 1712, the basic elements of the steam engine were brought together by the English blacksmith Thomas Newcomen. The Newcomen engine functioned by creating steam in a boiler, related to Papin's digester, and allowing the steam to fill a cylinder which had been fitted with a piston. When the steam was condensed a vacuum formed inside the cylinder and caused the weight of the atmosphere to force the piston to the bottom of the cylinder, creating the power stroke. Although the Newcomen engine was not fuel efficient (an obstacle that would be surmounted by James Watt in 1769), it spread quickly throughout England and was eventually exported to the rest of Europe.[49]

The steam engine was a revolutionary new device that forced people to think about the relationship between technology and the world in new ways. Unlike the earlier miner's friend, the Newcomen engine (and later the Watt engine) was a true prime mover that converted heat into work. Previously machines only transmitted power that existed in flowing streams, the blowing wind, or moving animals. The steam engine appeared to create its own motion. (By the nineteenth century, the science of thermodynamics would show that the steam engine was also only converting a natural source of power—heat—into mechanical work.) In any case the steam engine seemed to be fundamentally different from the older sources of power. Waterwheels and windmills could be understood in mechanical terms since they simply transmitted the motion of water or air by a series of mechanical linkages.[50] But during the eighteenth century there was no way to understand the action of the steam engine in such basic mechanical terms. As a prime mover, the steam engine seemed to many people to be based on a different model—one more closely associated with an organic world view. Just as the iatromechanist of the seventeenth century turned to mechanical models such as the pump to understand the new physiology of William Harvey, some scientists and technologists of the eighteenth century turned to physiology to understand the steam engine.

The most obvious model for a prime mover was vital activity. Throughout history, heat had been closely associated with vital phenomena and it was widely accepted that the source of organic activity was some process similar to combustion.[51] By the eighteenth century, several physiologists had focused on the problem of animal heat, that is, the process by which warm-blooded animals were able to keep a constant body temperature.[52] It was believed that the blood was heated either in the heart by a type of combustion or by the friction caused by its movement through the arteries and veins. This heat would continue to build up in the body if it were not for the cooling effect of breathing, which allowed the heat to be dissipated into the air. Therefore, the vital activity of the body, the source of its movement, depended on a process of alternative heating and cooling. The steam engine was based on a similar process; water was heated to form steam that was then condensed by cooling so as to form a vacuum which caused the motion of the piston to do work.

Although we tend to think of the steam engine as a prime example of mechanical technology, many of those associated with its early years saw it in organic terms. In 1725, Henry Beighton, an engineer closely associated with one of the first Newcomen engines, published a verse in the form of an enigma in *The Ladies' Diary*, a popular publication of the time.[53] Part of the verse states:

> My heart has ventricles, and twice three valves;
> Tho' but one ventricle, when made by Halves.
> My *Vena Cava*, from my further ends
> Sucks in, what upward my great Artery sends.
> The Ventricles receive my pallid blood,
> Alternate, and alternate yield the Flood:
> By Vulcan's Art my ample Belly's made;
> My Belly gives the Chyle with which I'm fed;
> From Neptune brought, prepar'd by Vulcan's aid.[54]

In the next volume of the *Diary*, Beighton revealed that his poem was a description of the invention of the steam engine. Although the verse is only one example, the fact that it was written by an engineer closely associated with the new steam engine and that it was published in a popular work, suggests that even in an age dominated by mechanical philosophy, some people saw the steam engine as a product of an organic world view.

The Machine as Plant

During the early nineteenth century, the Romantic philosophy, which criticized the machine as symbolizing the mechanical world view, also provided a model for a new approach to technology based on organic principles. A careful study of nineteenth-century technology shows that, contrary to popular opinion, a significant number of engineers were discovering new connections between the organic and the mechanical. Organic ideas did not dominate nineteenth-century technology, but such ideas did contribute to the development of what on the surface appears as a primarily mechanical view of technology.

One of the most influential organic ideas was the concept of classification. During the nineteenth century the development of a theory of mechanisms focused on this concept of classification.[55] In earlier periods it was common to consider every machine as a separate and distinct whole, consisting of parts peculiar to it. Machines were classified according to the purpose for which each machine was designed with little emphasis placed on underlying relationships between different types of machines.[56]

By the end of the eighteenth century, the search was on for a more systematic way to classify machines. A decisive step was taken by Gaspard Monge and his followers at the newly established *École polytechnique* in Paris. Monge argued that a study should be made of the elements of those machines which converted one type of motion into another. He said, "The most complicated machines being merely the result of a combination of some of these elements, it is necessary that complete enumeration of them should be drawn up."[57] The program of classification outlined by Monge and developed by his followers, particularly Jean Nicolas Pierre Hâchette and Agustin de Betancourt, was based on the conversion of one type of motion into another. Each motion was defined by two characteristics: one describing the type of motion (rectilinear, circular, or curvilinear) and the other the direction of motion (continuous or alternative). This led to classifying mechanisms into twenty-one possible categories (e.g., continuous rectilinear motion converted into continuous circular motion). The introduction of such a theory of classification of mechanisms closely paralleled similar schemes in natural history such as Linnaeus's binomial classification system. Just as naturalists had moved away from visible structure as a criterion of identity and toward an internal principle that would relate the visible to a deeper cause, engineers moved toward a similar organic classification system.

By the middle of the nineteenth century, there were attempts at establishing an even more "naturalistic" classification system based on the functional aspects of mechanisms. Instead of classifying in terms of the changes produced in the speed and direction of a given motion, Robert Willis (Jacksonian Professor at Cambridge) focused on the relationship of motions created by a mechanism. A clockhand mechanism was classified as a device that simply maintained the angular velocity of the hands in the ratio of twelve to one and kept their direction of rotation similar. The operation of the mechanism was independent of the given motion since it would still function if the motion was a back-and-forth rotation rather than a continuous one.[58] Willis's system could be considered an organic classification system since it linked superficial characteristics such as the mode of connection (rolling contacts, sliding contacts, gears, etc.) to essential functions such as directional and velocity ratios.

By the second half of the nineteenth century, theories of mechanisms were being debated in terms of organic models. For example, Franz Reuleaux, professor of engineering at Zurich and later director of the Royal Technical School of Charlottenburg, criticized earlier systems: "Monge's classification, however natural it appears, does not in the first place correspond to the real nature of the matter. Did it really do so—did it resemble, for instance, the classification of Linnaeus and Cuvier in organic nature—it would like them, be able to make its footing firm."[59] Reuleaux's theory was a movement away from individual isolated mechanisms and toward the idea of mechanisms as part of a continuous process or integrated system. Rather than treat a crank joined to a rocker arm by a connecting rod as a simple combination of three parts, Reuleaux perceived that the frame, which held the axle of the crank and the rocker in their positions, formed a fourth member in the combination, and the whole system consisted of a closed loop or circuit.[60] In such a circuit any member might be considered fixed and the others moving. These "inversions" would be other mechanisms related to the original. Reuleaux was in this way able to establish a new classification system based on the "natural" relationships that existed between mechanisms.

Organic Mechanisms

Many nineteenth-century engineers were even more explicit about their organic world view. Several works of the period argued that engineers should look to the natural world for inspiration for new inventions and designs. In 1842, Thomas Ewbank (U.S. Commissioner of Patents, 1849–52), noted, "Few classes of men are more interested in studying natural history, and particularly the structure, habits, and movements of animals, than mechanics; and none can reap a richer reward for the time and labor expended upon it."[61] Ewbank traced many of the nineteenth-century developments in technology to some type of organic process:

> The flexible water-mains . . . by which Watt conveyed fresh water under the river Clyde were suggested by the mechanism of a *lobster's tail*—the process of tunneling by which Brunel has formed a passage under the Thames occurred to him by witnessing the operations of the *Teredo*, a testaceous *worm* covered with a cylindrical shell, which eats its way through the hardest wood—and Smeaton, in seeking the form best adapted to impart stability to the light-house on the Eddystone rocks, imitated the contour of the bole of a *tree*.[62]

One of the strongest advocates for an organic approach to technology was George Wilson, the first professor of technology at the University of Edinburgh and director of the Industrial Museum of Scotland.[63] The basic thrust of Wilson's ideas can be gathered from his inaugural lecture, delivered in 1855, "What Is Technology?" and his introductory lecture of 1856 "On the Physical Sciences Which Form the Basis of Technology."[64] In the inaugural lecture, he defined the object of technology as the discovery of ''the principles which guide or underlie Art,''[65] and he implied that such principles were to be found in the organic world rather than the mechanical world. He said, ''The mason-wasp, the carpenter-bee, the mining caterpillars, the quarrying sea-slugs, execute their work in a way which we cannot rival or excel. The bird is an exquisite architect; the beaver a most skillful bridge-builder; the silk-worm the most beautiful of weavers; the spider the best of netmakers.''[66] Although he recognized the differences between animal instincts and conscious human invention, animal instincts seemed to have the implicit character of underlying principles for which Wilson was searching.

The next year Wilson attempted to outline the scientific basis of technology and in so doing established an essential connection between technology and the organic. He began by admitting that almost all of the sciences could have some influence on technology but argued that technology depended on each science in a different way. He divided the sciences that formed the basis of technology into two groups.[67] Sciences such as astronomy, geology, meteorology, and mineralogy were observational and classificatory since they influenced technology by discovering natural phenomena such as the motion of the planets, atmospheric changes, and mineral and vegetable products which could be of practical use. Sciences such as chemistry, mechanics, heat, electricity, and magnetism were transformational and directive since they influenced technology by transforming materials discovered by the observational sciences.

Although each of these sciences contributed something to technology in its own way, Wilson realized that the real advancement of technology required a synthesis of the observational and the transformational. Such a synthesis was achieved in the one science missing from Wilson's classification; this ''remarkable science'' was none other than biology, which, being both observational and transformational, registrive and directive, naturalistic and experimental, could be the true basis of technology.[68]

For Wilson, the organism became the primary archetype of technology since its function was to take a raw material and transform it into a finished product. ''The plants and animals which as agriculturalists we care for, may be regarded as skilled labourers, who in return for food, wages . . . and a certain liberty of action, agree to collect or manufacture for us a multitude of useful substances.''[69] Wilson believed that organic nature could provide engineers with the basic designs needed for new inventions. He argued that the industrialists ''may imitate, in whole or in part, vegetable or animal organisms (including animal workmanship), accepted as divine patterns ready to his hand.''[70] A specific example that he noted was the ''study of the torpedo [electric eel] that enabled electricians to understand some of the most important laws of action of their artificial machines and batteries'' so that ''inorganic electricity, both as a science and an art, is very largely indebted to organic electricity, alike for the explanation of the laws which it obeys, and for the contrivances by which it works.''[71]

But Wilson went beyond suggesting that engineers imitate nature. He believed that technology, at its foundation, was organic. For an organically based technology, he suggested a framework radically different from that imposed by a mechanical view of technology. In Wilson's system, machines would be seen as alive: "There is . . . the fact that, to the extent an organism can be wielded by us, it enables us to add to the transforming and transmuting powers of mechanical and chemical force, which alone are available to the dead machine, the metamorphosing power of vital force."[72] Wilson was trying to create a technological theory based not on mechanical force but on vital force. He argued that actual machines embodied the living force that went into producing them: "We do not sufficiently remember that all other machines are the offspring of living machines. A steam-engine is the literal as well as the metaphorical embodiment of so much horse-power. A railway viaduct is the petrifaction of so much animal force."[73] For Wilson, the fundamental forces of science were vital, and even mechanical forces were simply a different manifestation of vital force. He believed technological machines were a form of domesticated living force.

Technology and Evolution

By the second half of the nineteenth century, evolutionary ideas began to be associated with technology.[74] If machines and tools were based on organic principles, it seemed likely that the development of machines and tools should be governed by the same evolutionary forces that were at work in nature. In his book *First Principles,* Herbert Spencer argued that technology, like the rest of nature, was governed by evolutionary forces. He said, "The progress from small and simple tools, to complex and larger machines, is a progress in integration."[75] Through the process of evolution, simple elements were synthesized together to form a new whole that was greater than the sum of its parts. Spencer stated, "A modern apparatus for spinning or weaving, for making stockings or lace, contains not simply a lever, an inclined plane, a screw, a wheel-and-axle, joined together, but several of each—all made into a whole."[76]

The most important evolutionary theory of tools reflected the new interest in anthropology, but many of the leading figures in this field had backgrounds in biology or engineering.[77] One of the most influential figures in this movement was Lieutenant-General A. Lane-Fox Pitt-Rivers, whose collection of tools forms the basis of the ethnographic museum in Oxford.[78] Combining his work in the military with his interest in the ideas of Darwin and Spencer, Pitt-Rivers began to apply the idea of evolution to machines and tools. While studying the historical developments of firearms, he began to notice how they had been improved through small successive changes that paralleled organic evolution, and during the 1860s he lectured at the Royal United Service Institution on the evolution of weapons. According to Henry Balfour, first curator of the Pitt-Rivers Museum, "Through noticing the unfailing regularity of this process of gradual *evolution* in the case of firearms, he was led to believe that the same principles must probably govern the development of the other arts, appliances and ideas of mankind."[79]

For simplicity, Pitt-Rivers focused his collection on primitive artifacts and tools, but he made it clear that the ideas could be applied to modern technology.[80] Through the study of primitive artifacts, Pitt-Rivers and Balfour were able to show that "apparently quite separate and distinct implements, of quite different appearance and function, could in fact be 'generically' related through a whole number of transitional forms."[81] Similar to the problem of missing links in Darwinian evolution, many of the transitional forms of artifacts may have disappeared, giving the impression that different objects were independent of one another. This was especially true of modern inventions where many times the intermediate versions existed only in the inventor's mind or in experimental form.[82]

Pitt-Rivers and Balfour emphasized a concept of technological evolution in which both the form and the function of a particular tool or artifact could be placed in an organic relationship to other objects.[83] In this theory, the process of design or invention takes on the characteristics of a botanist's method in trying to develop a new hybrid plant. The inventor must constantly be aware of the "genetic" relationships that exist between machines. A single machine or tool does not exist by itself but as part of an organic system.

Technology, Organicism, and Karl Marx

One of the most influential examples of an organic theory of technology is found in the works of Karl Marx. Known mostly for his economic and political theories, Marx was also a historian and philosopher of technology, and his theory of technology played a central role in his major work *Das Kapital*. Marx had a lifelong interest in the philosophy of nature beginning in 1841 with his Ph.D. dissertation of the *Difference Between the Democritean and Epicurean Philosophy of Nature*, which focused on the theory of atomism, and continuing through his great admiration for Darwin's *On the Origin of Species*. In preparation for *Das Kapital*, Marx attended a course on technology taught by Professor Robert Willis, one of the people responsible for the establishment of a theory of classification of mechanisms.[84]

Marx saw a close connection between technology and the organic. At the beginning of his chapter on "Machinery and Modern Industry," in *Das Kapital*, he noted,

A critical history of technology would show how little any of the inventions of the 18th century are the work of a single individual. Hitherto there is no such book. Darwin has interested us in the history of Nature's Technology, *i.e.*, in the formation of the organs of plants and animals, which organs serve as instruments of production for sustaining life. Does not the history of the productive organs of man, or organs that are the material basis of all social organisation, deserve equal attention?[85]

It was Marx's purpose in *Das Kapital* to provide such a study of man's "productive organs." He believed that the "relics of bygone instruments of labor possess the same importance for the investigation of extinct economic forms of society, as do fossil bones for the determination of extinct species of animals."[86] For Marx, these "instruments of labor" were, in fact, organic systems. He argued that the instruments "of a mechanical nature, taken as a whole, we may call the bones and muscles of production," while such instruments as pipes, tubs, and jars, which

Greek automaton in which the two figures pour wine and the serpent hisses when a fire is lit on the altar. From *The Pneumatics of Hero of Alexandria*, trans. and ed. by Bennet Woodcroft (1851).

Second astronomical clock in Strasbourg Cathedral (completed in 1574), containing both a model of the universe and small mechanical figures. From a woodcut by Tobias Stimmer in "Eigentliche Fuerbildung und Beschreibung dehs newen Kuenstrechen Astronomischen Urwerks," Strasbourg, 1574.

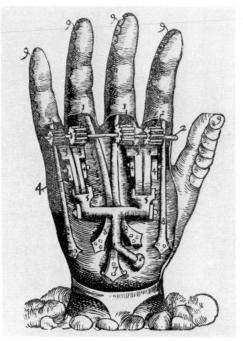

Mechanical hand. From Ambroise Paré, *Les Oeuvres d'Ambroise Paré*, Paris: Gabriel Buon, 1579.

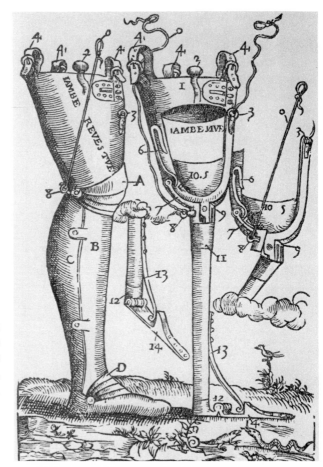

Artificial legs for knights on horseback. From Ambroise Paré, *Les Oeuvres d'Ambroise Paré*, Paris: Gabriel Buon, 1579.

Mechanical singing bird powered by water. From Robert Fludd, *Utiusque Cosmi Historia* (1618).

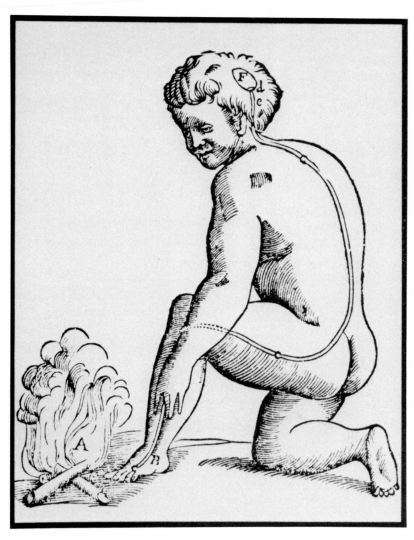

Cartesian model of the human body. Fire causing the expansion of nervous fluid, resulting in an involuntary motion away from the heat. From *René Descartes,* Treatise of Man (1662).

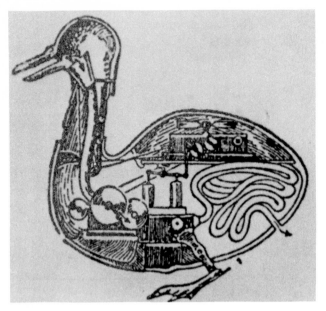

Detail of Jacques Vaucanson's duck, showing the internal mechanisms for digestion. From A. Chapuis and E. Gelis, *Le Monde des Automates,* Neuchatel: Editions du Griffon, 1928.

Experimental apparatus and machines for the study of animal electricity. From Luigi Galvani, *De Viribus Electricitatis in Motu Musculari* (1791).

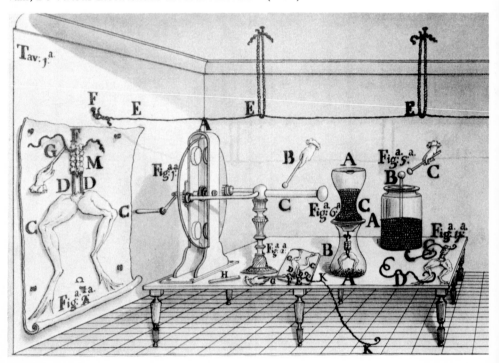

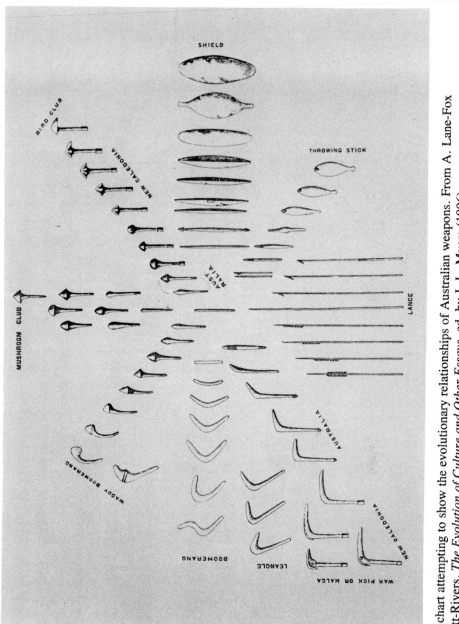

A chart attempting to show the evolutionary relationships of Australian weapons. From A. Lane-Fox Pitt-Rivers, *The Evolution of Culture and Other Essays*, ed. by J. L. Myers (1906).

The Harvard-IBM Mark I computer, 1944. One of the earliest computers to function as an information processing device. The electrically powered machine performed calculations through a series of mechanical relay switches and wheels. Courtesy of International Business Machines Corporation.

A step toward a nanotechnology. This miniature motor is smaller than the width of a human hair and is powered by electrostatic energy. Such motors might be able to perform microsurgery by functioning inside human blood vessels. (A 130 μm-diameter electrostatic micromotor with 1.5 μm air gaps. Courtesy Mehran Mehregany, Massachusetts Institute of Technology.)

"hold the materials for production . . . we may in a general way, call the vascular system of production."[87]

A central point of Marx's criticism of the capitalistic system is that modern industry had taken the tools away from the craftsmen; by incorporating those tools into a modern machine, capitalism had reduced the skilled craftsmen to simple workers. This criticism rests on an organic view of technology. For Marx, the primary reason that machines could take over the work of craftsmen was that machines were, in some sense, alive. He said,

> An organised system of machines, to which motion is communicated by the transmitting mechanism from a central automaton, is the most developed form of production by machinery. Here we have, in the place of the isolated machine, a mechanical monster whose body fills whole factories, and whose demon powers, at first veiled under the slow and measured motions of his giant limbs, at length breaks out into the fast and furious whirl of his countless working organs.[88]

For Marx, the epitome of modern technological production was the factory with its system of automated machines. In fact, it was through the factory system that technology most thoroughly manifested its organic form. In one of his descriptions of the automatic factory, Marx quotes Dr. Andrew Ure, an early nineteenth-century scientist and author of *The Philosophy of Manufacturers* (1835), who called the factory "a vast automaton, composed of various mechanical and intellectual organs, acting in uninterrupted concert for the production of a common object, all of them being subordinate to a self-regulating moving force."[89] In this factory Marx believed that "the workmen are merely conscious organs, [and] co-ordinate with the unconscious organs of the automaton."[90] The workers' relationship to the machines was also seen in organic terms, "Among the attendants [of the machines] are reckoned more or less all 'Feeders' who supply the machines with the material to be worked."[91] Marx was not using organic terminology simply as a metaphor for the machine. It was precisely because the machines of modern industry incorporated actual vital or living force that they were considered a threat to the working class.

Organic Structures

Throughout the eighteenth and nineteenth centuries, and especially in the second half of the nineteenth century, the organic world view, not only influenced the development of machines, but had a significant impact on the development of structures.[92] Organic theories provided engineers and architects a model for understanding structures. In 1770 Jean-Rodolphe Perronet, a civil engineer and professor at the *École des Ponts et Chaussées,* argued that the structural elements of buildings imitated the structure of animals. He noted that in Gothic cathedrals, "the high, delicate columns, the tracery with transverse ribs, diagonal ribs and tiercerons, could be compared to the bones, and the small stones and voussoirs, only four or five inches thick, to the skin of these animals."[93] In 1801 J. N. L.

Durand, a writer on architecture, attempted to classify buildings according to their function, the same way a botanist such as Goethe had classified plants.[94] He believed this classification system would allow students to understand the underlying principles of building. During the 1850s, the famous architectural theorist, Eugene Emmanuel Viollet-le-Duc concluded from his study of medieval cathedrals that masons "sought to bring out in the structures of their buildings those qualities they found in vegetation."[95]

Organic theories also provided engineers and architects a repository of designs that could be used in structures. As early as 1756 engineer John Smeaton used the shape of a tree-trunk as a model for the lighthouse he was designing for the Eddystone rocks off the southern coast of England.[96] Since an oak tree was able to resist great winds without being uprooted, Smeaton argued that its shape might be appropriate for a lighthouse on a stormy site like the Eddystone rocks. A trunk's broad base had a low center of gravity, giving stability, while its narrowed upper part reduced resistance to wind. In 1850 Joseph Paxton, a superintendent of gardens for the Duke of Devonshire, submitted the winning design for a building to house the Great Exhibition of 1851 in London.[97] His revolutionary design for a Crystal Palace, featuring a glass roof with flanged undersurfaces, was inspired by the *Victoria Regia* water lily.

By 1863, the editor of the *Revue Générale de Architecture* was calling for an "organic architecture." The first architect to specifically design organic architecture was Louis Sullivan, one of the first modern architects and leader of what became known as the Chicago School.[98] He had read many of the nineteenth-century biologists, including Darwin, T. H. Huxley, Asa Gray, and Herbert Spencer, who particularly influenced him. Central to Sullivan's organic architecture was the idea that "form follows function." Function was not simply structural, but involved social factors as well; a building evolved naturally, governed by both structural and social forces. Sullivan's ideas were passed on to his protegé, Frank Lloyd Wright, who wrote, "By organic architecture I mean an architecture that develops from within outward in harmony with the conditions of its being as distinguished from one that is applied from without."[99]

Conclusions

By the middle of the nineteenth century, the organic world view had expanded to include technology within its realm of explanation. Although organic theories did not dominate technology, they did provide a significant alternative to the mechanical view of technology, and they play an important role in our study of the relationship between technology and organic life. Within the organic world view it was possible to reconcile the apparent dichotomy between the machine and nature. Believers in this world view would argue that all forces active in the world, even those harnessed in technology, are at their foundation vital forces and therefore machines must incorporate some level of vitality. The apparent conflict between the machine and nature is resolved by encompassing technology within the organic view. Since machines were, in fact, alive, it might even be possible for them to have developed through a process of organic evolution.

But several problems with organic theories of technology remained. First, although organic theories of machines could explain technological processes, the exact relationship between the vital principle and the technological object was vague. It was not clear whether machines contained an autonomous vital force, or whether this vitality simply emerged out of the organization of the elements of a machine.

Second, the organic theory of machines, like mechanical philosophy, also had a problem with the relationship between spirit and matter. The organic world view argued that the material world was not inert and passive; it was infused with some kind of vital force. Clearly there is some distinction between a machine and a product of nature, but the organic view of technology never made it clear what that distinction should be.

Finally, the organic theory of vitalism was being questioned during the second half of the nineteenth century. Many people still clung to mechanical philosophy, but a more common attitude was one of ambivalence or agnosticism. For a significant number of people neither the mechanical nor the organic world view continued to make sense.[100]

Nineteenth-century discoveries like the concept of cells and the concept of energy, and subsequent twentieth-century developments in genetics and quantum mechanics could not be easily explained by either the mechanical or the organic world view. A new world view was beginning to emerge, one that went beyond the limitations of the previous world views and one that would establish a fundamentally new concept of technology and organic life.

6

The Emergence of
the Bionic World View

Science is taking on a new aspect which is neither purely physical
nor purely biological.

ALFRED NORTH WHITEHEAD, *Science and the Modern World*

By the middle of the nineteenth century, distinctions between the mechanical and
organic world views had become blurred and confused, and the integrity of each
view threatened by developments and changes. With the introduction of the
Newtonian concept of force, supporters of mechanical philosophy had continual
problems explaining force in terms of matter and motion, and had to resort to
accepting a new active principle as part of the mechanical world view. At the same
time, some of the successes of mechanical philosophy in the area of physiology
raised questions concerning the idea of vital force, while supporters of organic
philosophy began to accept the notion that vital force was an emergent property of
matter similar to Newtonian force. Darwinian evolution also called into question
fundamental assumptions of both world views. The idea of chance variation,
which was at the center of Darwin's theory, undermined both the idea of organic
teleology and mechanistic determinism.[1] Finally, nineteenth-century developments such as cell theory and electromagnetic field theory supported aspects of
both world views. It was no longer clear that either view could explain the world.

One of the best popular expressions of this ambiguous situation was Samuel
Butler's satirical utopian novel *Erewhon* (1872).[2] Butler, who had a lifelong
interest in natural philosophy, was especially interested in the debate between the
Darwinists, like T. H. Huxley, who held to a mechanistic explanation of
evolution, and the Lamarckians, who held to a vitalistic interpretation of organic
evolution. In his youth Butler had been an early supporter of Darwinian evolution
but after discovering the alternative Lamarckian view, Butler became converted to
a more vitalistic interpretation of evolution.[3] By 1877, he supported the
Lamarckian theory to the extent of writing a scientific treatise *Life and Habit*,
which attacked the mechanistic and materialistic basis of Darwinian natural
selection and put forward in its place a vitalistic theory of biology.

Although he revised the final version of *Erewhon* in 1901 to reflect his support of vitalism, his original 1872 version left unresolved the question of mechanism versus vitalism. As a recent scholar notes, "What bothered even Butler's contemporaries about *Erewhon* was their uncertainty about what was being satirized and what position the author was taking. . . . [T]he genius of the 1872 *Erewhon* is . . . in attitudes that attach themselves to any side of a question or to many sides at the same time."[4] The ambivalence in Butler's thought can be seen in the section of the novel where the hero visits the Colleges of Unreason and hears about the Civil War between the machinists and the anti-machinists as chronicled in "The Book of Machines." Here Butler presents the debate between two professors concerning the relationship between machines and life. At the center of the debate is the question whether machines are to be considered as modeled on organic life or whether human beings are to be considered as modeled on machines.

The first professor makes an argument that Butler means as a satire of the biologist who used that machine as a model for life.[5] In order to support his argument that machines can evolve and threaten human beings, the first professor argues that organic life is essentially mechanical. He says:

> If it be urged that the action of the potato is chemical and mechanical only, and that it is due to the chemical and mechanical effects of light and heat, the answer would seem to lie in an enquiry whether every sensation is not chemical and mechanical in its operation? . . . Whether strictly speaking we should not ask what kind of levers a man is made of rather than what is his temperament?[6]

This line of reasoning leaves room for hardly any other explanations. According to this professor, "Man's very soul is due to the machines; it is a machine-made thing: he thinks as he thinks and feels as he feels through the work that machines have wrought upon him."[7] If humans are simply machines and if humans have evolved from lower forms of life through a mechanical process of evolution, the first professor argues, machines too must be able to undergo a process of evolution that will make them superior to humans. He says,

> The vapour-engine must be fed with food and consume it by fire even as man consumes it; it supports its combustion by air as man supports it; it has a pulse and circulation as man has. It may be granted that man's body is as yet the more versatile of the two, but then man's body is an older thing; give the steam-engine but half the time that man has had, give it also a continuance of our present infatuation, and what may it not ere long attain to?[8]

Here, satirically, Butler criticizes the mechanical view of life by showing that its application could lead to an unforeseen conclusion. If organic life is simply mechanical, then we are forced to accept machines as equivalent to organisms, and therefore machines must be just as "alive" as organisms.

But Butler also presents the arguments of a second professor who represents the vitalistic position of biologists such as Lamarck. To support his position that machines were not gaining dominance over humans, this character argues that machines were the means by which humans were themselves evolving. Here the model is organic rather than mechanical. According to the professor,

> machines were to be regarded as a part of man's own physical nature, being really nothing but extra-corporeal limbs. . . . The lower animals keep all their limbs at home in their own bodies, but many of man's are loose, and lie about detached, now here and now there, in various parts of the world. . . . A machine is merely a supplementary limb; this is the be all and end all of machinery.[9]

Unlike the first professor's view in which machines become alive and take over the world, here the danger is that humans will regress. The second professor worries that the dependence on machines "might cause a degeneracy of the human race, and indeed that the whole body might become purely rudimentary, that man himself being nothing but soul and mechanism, an intelligent but passionless principle of mechanical action."[10] If machines are thought to be some external projection of the vital spirit of the human mind, then we may be forced to accept the mind as a "passionless principle of mechanical action," and humans as a "machinate mammal."[11]

Although Butler later would come to intend "The Book of Machines" as a criticism of the mechanistic approach to biology, in 1872 he was entirely ambivalent.[12] There are no indications that the vitalistic position is to be taken any more seriously than the mechanistic. Thus "The Book of Machines" represents the confusion that existed in the late nineteenth century over the relationship between technology and organic life, when neither the mechanical nor the organic world view seemed satisfactory. What was needed was a new model to incorporate both technology and the organic. The origins of this new world view can be seen in the transformation that took place during the nineteenth century in both the theory of organisms and the theory of matter. What emerged was a model of the world that was in one sense *neither* mechanical nor organic, and in another sense *both* mechanical and organic.

Teleomechanism

Much of nineteenth-century biology revolved around the debate between the vitalists, who believed that some distinctive, purposeful force guided the development of living organisms, distinguishing them from all other types of matter; and the mechanists, materialists, or reductionists, who believed that life could be reduced to some physical or chemical phenomena.[13] But as Timothy Lenoir has shown in *The Strategy of Life*, in the first half of the nineteenth century, particularly in Germany, there emerged a research program in biology which combined what most biologists had assumed to be contradictory ideas—teleology and mechanism.[14] This program, called teleomechanism, had its basis in the philosophical writings of Immanuel Kant, who attempted to limit biological explanations to mechanical principles while admitting the unique character of organic phenomena. Kant argued that, based on experience, the purposeful organization of biological phenomena must be a given that cannot be explained by mechanical principles. But once this purposeful organization is accepted, the role of mechanical processes within an organism could be investigated. For many of the teleomechanists, the organizing principle of life, or *Bildungstrieb,* emerged from a particular arrangement of mechanical processes, but mechanical theories could not explain the origins of that particular arrangement. Although this approach had connections to vitalism, it was a "vital materialism."[15]

This new teleomechanism came at a time when more traditional vitalistic theories were coming under attack. In 1814 the chemist J. J. Berzelius demonstrated that many organic substances obeyed the same laws of chemical combination as did inorganic substances.[16] In 1828 another German chemist, Friedrich Wohler was able to synthesize urea, thought to be producible only in living organisms, through normal laboratory chemistry.[17] Although Wohler's work did not eliminate the distinction between organic and inorganic, it did suggest that living materials were simply more complex arrangements of normal inorganic substances. By the middle of the century, the emphasis in vitalism shifted away from the idea of a vital substance and toward the concept of special laws of vitality or special vital forces that acted on the normal chemical substances, and made those substances combine and act in ways that were unique to organic materials. Thus, living organisms were composed of the same materials as inorganic substances, but aside from the normal physical and chemical forces that governed the actions of the materials in inorganic bodies, they were also subject to the actions of vital forces or laws. These vital forces might be related to physical and chemical laws but they also differed from those laws in some way.

One theory about the difference, reflecting the ideas of teleomechanism, was put forth by the German physiologist Johannes Müller. He believed that atoms normally combined in pairs when forming chemical compounds while organic molecules were composed of three, four, or more elements.[18] That is, organic systems contained the same chemicals as nonorganic systems, but these chemicals were combined together in special ways. For Müller, it was the existence of a *vis essentialis* (vital principle or life force) that allowed normal atoms to organize themselves into organisms. Like gravity, this force could not in itself be observed; rather it could be known only through its actions on matter. In fact, Müller and other vitalists remained vague as to the exact nature of the vital force.

A scientist who tired to be more specific concerning the vital force was Justus Leibig. Although he believed that organisms were formed and maintained by a special life force (*Lebenskraft*), he did hold that "the only known ultimate cause of the vital force within plants or animals is a chemical process."[19] Leibig suggested that when normal physical and chemical forces were brought together in living substances, the life force emerged as a consequence of the complex organization of that material. Although the vital force was still distinguished from normal physical and chemical forces by such properties as its capacity for self-propagation, it seemed related to those forces.[20]

Cell Theory

At the center of the new program of teleomechanism was the development of cell theory or cytology.[21] As early as 1664 there are indications that scientists were using the microscope to observe the cellular structure in plants. Robert Hooke noted the "Honey-comb" texture of petrified wood, but these early observations had little effect on the theoretical development of biology. Most eighteenth-century physiologists focused on the concept of organs or "tissues" rather than on "cells." Nathan Grew observed cells in plants but thought of them as a "fine fabric or tissue."[22] When cells were identified they were not seen as a universal phenomenon but only as a particular type of tissue. For example, the eighteenth-century Swiss physician Albrecht von Haller classified tissues into three categories—nervous, muscular, and cellular.[23] This last category comprised connective tissues that appeared to be cell-like because of their fibrous quality. Because it was generally believed that fibers were the source of vital activity, cells were considered to be cavities between fibers rather than entities themselves.[24]

Before the nineteenth century the development of a cell theory was impeded by the limitations of microscopy. Most of the early microscopes were not powerful enough to observe animal cells and had difficulty discerning a structure inside plant cells. But some theoretical support for a more complex concept of cells arose from the study of embryology. Many of the eighteenth-century supporters of epigenesis argued that a cell must be something more complex than simply an empty cavity. The German biologist C. F. Wolff held that plants at all stages of their development were composed of small liquid-filled bladders or "vesicles," which arose from some undifferentiated fluid. And in 1805 the *naturphilosopher* Lorenz Oken wrote, "The primitive mucous vesicle is called an *infusorium*. Every organism is a synthesis out of infusoria. Organic development consists of nothing but the accumulation of an unlimited number of mucous particles or infusoria."[25]

By the 1830s the basic elements of a cell theory began to emerge. Much of the credit belongs to the German biologists Matthias Jacob Schleiden and Theodore Schwann, both of whom were influenced by teleomechanism.[26] Based on microscopic studies of plants, Schleiden proposed the argument in 1838 that all plants were composed of cells. In doing so he noted one of the key elements of a biological theory based on cells: "Every cell leads a two-fold life; one quite independent and belonging to its own individual development, the other dependent to the extent that cells become integrative parts of plants."[27] It was this dual nature of cells that would connect cell theory to a tradition of teleomechanism. The mechanists could focus on the cell as an independent entity while the vitalists could emphasize the cell's role as part of a holistic system. By incorporating both mechanical and vitalistic elements, cell theory would make a significant contribution to a new world view.

The cellular structure of animals was much harder to observe, but through conversations with Schleiden, Schwann began to suspect that cell theory could be extended to animals. Although he could not see a cellular structure in all animal tissue, Schwann did observe that all of the tissues contained a body that could be associated with the nucleus of a plant cell.[28] By 1839 Schwann was ready to argue that both plants and animals were composed of cells. In doing so, he was formulating a new foundation for the organic world. For Schwann, the cell was the center of vital activity: "The universality of respiration is based entirely upon . . . the metabolic phenomena of the cells."[29]

Surprisingly the weakest part of the Schleiden–Schwann theory was their hypothesis of cell formation. According to them, cells were formed around the dark granule, or "nucleous," by a process similar to crystallization. For Schleiden and Schwann, cells did not arise from other cells; they developed out of a structureless liquid or *cytoklastema* and then attached themselves to other cells. By making cell production analogous to crystallization, they hoped to be able to explain cell formation in terms of mechanical principles. Although Schleiden and Schwann had established cells as the basic units of plants and animals, their notion that cells were mechanically formed out of a nonliving liquid quickly came under attack. During the 1840s and 1850s new advances in microscope design allowed scientists to observe cells being formed by the division of existing cells. With the idea of cell formation through division (mitosis), cell theory continued its association with a vitalistic tradition. Although cells could be considered as the mechanical units that composed an organism, these organisms could arise from other organisms only through the self-replication of the cell.

After Schleiden and Schwann, the vitalistic and mechanical properties of the cell were widely debated. During the middle of the century the debate focused on the actual composition of the cell. The development of cell theory had again raised the question of the source of life in an organism.[30] The supporters of vitalism attributed life to a special vital substance that composed the cell. By the middle of the century, vitalists were describing this as a universal living jelly, or protoplasm, a unique material common to all life.[31] But for the mechanists, the source of life was simply the metabolic processes of the cell which were functioning according to the normal laws of physics and chemistry. For them, the content of the cell was a very complex material, or cytoplasm, that could be ultimately reduced to normal chemicals.

A real synthesis of the vitalist and reductionist views began to be realized in the mid-century work of the German pathologist Rudolph Virchow. In 1858 he put forward the idea of a "new vitalism."[32] In an attempt to reconcile vitalism and mechanism, Virchow emphasized, on the one hand, the cell as the fundamental unit of organic matter, and, on the other hand, life as the sum total of the actions of individual cells. That is, he argued that life was "a mechanical phenomenon in which ordinary matter obedient to ordinary laws operates under highly extraordinary conditions."[33] The cells were independent units whose actions together resulted in the phenomenon of life. Life was distinguished from other mechanical processes because cells acting in concert formed a unified whole. An inorganic material could be formed into a crystal through the mechanical action of physical and chemical forces, and those forces would form a crystal whether or not that particular material had been a crystal before. But a living system was formed in a different way. According to Virchow, a cell could not arise from the actions of a group of molecules or atoms unless they were already organized into a living system. Every living cell came from another cell (*omnis cellula ex cellula*).[34]

Virchow's life force (*Lebenskraft*) is both vital and mechanical. It is mechanical in the sense that it is simply the product of the organization of matter, but vital in the sense that it produces that organization. When a group of molecular forces come together in a certain way their actions as a whole differ dramatically from the sum of their individual actions. As Virchow said:

> Nowhere have I even hinted that life-force is primary, or specifically different from the other natural forces; on the contrary, I have repeatedly and emphatically stated the likelihood of its mechanical origin. [But] it is high time that we give up the scientific prudery of regarding living processes as nothing more than the mechanical results of molecular forces inherent in the constitutive particles.[35]

Physical Vitalism

The debates between the vitalists and the mechanists continued into the second half of the nineteenth century. Several physiologists continued to attack the idea of any kind of vital force, attempting to show instead that organic processes could be reduced to the physical and chemical laws that governed matter, motion, and force. Ironically some of the leading reductionists, such as Emil DuBois-Reymond and Hermann von Helmholtz, had studied in the Berlin laboratory of the vitalist Johannes Müller. Both DuBois-Reymond and Helmholtz began to apply the new concept of the conservation of energy to organic systems.[36] Dubois-Reymond argued that there was no way for a vital force to emerge from material phenomena without violating the conservation of energy. At about the same time, Helmholtz conducted experiments measuring the heat generated by the action of the muscles which showed that such heat arose from chemical processes within the muscle and not from any vital force.

But reductionism faced as many problems as did vitalism. On the one hand, vitalism was becoming a "theory of the gaps," filling in with its vague principle whatever could not be explained by mechanical principles. Reductionism, on the other hand, was more than able to fill in such gaps. For the reductionists, the ideal goal of biology was to explain life processes through the laws of physics and chemistry, but this attempt remained as much a goal as a reality. Life processes were found to be much more complex than the reductionists had originally anticipated. In 1872 the reductionist DuBois-Reymond admitted that the application of mechanical philosophy to biology would always be limited.

> With respect to the riddle of the corporeal world the scientist had long been used to avowing, with manly resignation, that he simply does not know (*ignoramus*). In looking back over the victory-strewn way he has come, he is borne up by the quiet awareness that what he does not know today he can possibly know in the future and that perhaps one day he will. But with respect to the other riddles—what matter and force are and how they can think—the scientist must take his stand for once and for all on the truth much harder to acknowledge, namely that, here, he never will know (*ignorabimus*).[37]

The teleomechanist program had attempted a reconciliation of vitalism and mechanism by accepting aspects of both; another solution to the problem was to reject both models and move to a new method that would transcend either. The great French physiologist Claude Bernard found weaknesses in both the vitalistic and the reductionistic points of view.[38] For example, he saw "at the basis of *vitalist doctrines* an irremediable error which is that the vitalist by a personification of the arrangement of things, is seduced into regarding that arrangement as a force and thus gives real existence, and material and efficacious action, to something immaterial which is in fact a production of the mind."[39] But there were equal problems in the reductionists' approach. He said:

> Admitting that vital phenomena rest upon physico-chemical activities, which is the truth, the essence of the problem is not thereby cleared up. . . . Vital phenomena possess indeed their rigorously determined physico-chemical conditions, but, at the same time, they subordinate themselves and succeed one another in a pattern and according to a law which pre-exist. . . . It is as if there existed a pre-established design of each organism and each organ such that, though considered separately, each physiological process is dependent upon the general forces of nature, yet taken in relation with the other physiological processes, it reveals a special bond and seems directed by some invisible guide in the path which it follows and towards the position which it occupies.[40]

Bernard's solution to the reductionist–vitalist controversy was to focus again on the cell. Organisms were not characterized by some vital principle but were distinguished from other forms of matter by the fact that the elements of a living body (the cells) were interconnected and mutually responsive to external changes. The controlling factor of this interconnectedness was the internal environment (*milieu intérieur*) in which the cells existed. This environment, the complex fluid in plants and the blood plasma in animals, was responsible for the stability of the organism and, further, could be studied in terms of physical and chemical laws. Bernard thus believed that the "conditions" of life could be analyzed by physical and chemical laws but that this did not mean that life itself could be reduced to such mechanical principles.[41] He called this approach "physical vitalism," meaning that physical and chemical processes could manifest themselves through their organization as vital phenomena.[42] In the end, Bernard's approach to the question of the definition of life was positivistic. Life could be analyzed in terms of mechanism or vitalism but the ultimate nature of life was an unanswerable question. As Bernard said, "To sum up our thought metaphorically, the vital force directs phenomena which it does not produce; and physical agents produce phenomena which they do not direct."[43]

Organicism and Holistic Materialism

By the end of the nineteenth century biology was becoming transformed. The previous mechanical and organic world views were no longer sufficient to explain organic processes. As historian Everett Mendelsohn has noted, the terms mechanism and vitalism "had lost their usefulness so far as providing an adequate classificatory system for nineteenth-century physiologists."[44] But there were still attempts at keeping the mechanist–vitalist debate alive.[45] In 1888 Wilhelm Roux, a leading embryologist, published the results of some experiments on frogs' eggs that seemed to support a mechanistic theory of biology.[46] When a frog embryo had reached the two-cell stage, Roux punctured and killed one of the cells using a hot needle. He discovered that the subsequent development of the embryo was abnormal and sometimes it failed to develop at all beyond a certain stage. He interpreted this experiment as supporting his "mosaic theory," in which each cell that arose from the division of another cell received different hereditary material, and eventually each cell in the fully developed organism would have a single unique potential. Killing one cell had resulted in abnormal development since it removed some hereditary material that existed uniquely in that cell, just as removal of one part of a machine would result in its incorrect functioning.

By the 1890s, Roux had used his work in embryology as the basis for a general mechanistic approach to biology that he labeled *Entwicklungsmechanik* or "developmental mechanics." Through this new program, Roux hoped to be able to explain biological phenomena in terms of physical and chemical causes.[47] A high point of the *Entwicklungsmechanik* approach to biology was the result of the work of the physiologist Jacques Loeb, who served as the model for Max Gottlieb in Sinclair Lewis's *Arrowsmith*.[48] In his book *The Mechanistic Conception of Life* (1912), Loeb summarized his earlier biological research in which he tried to create "a technology of living substance."[49] Loeb's most spectacular evidence for a purely physical-chemical approach to biology was his discovery in 1899 that the unfertilized eggs of sea urchins could be made to begin developing through purely physical manipulation, such as pricking them with a needle, or purely chemical manipulation, such as altering the salinity of the sea water in their culture. His work attracted tremendous publicity in the popular press, where it was used to support everything from the doctrine of the Virgin birth to warnings that "maiden ladies" should give up bathing in the ocean.[50] But to some biologists, the discovery of artificial parthenogenesis indicated that the basis of life could be understood in terms of physical chemistry.

Although biologists such as Roux and Loeb still saw biology as a battleground between mechanism and vitalism, other biologists saw problems with both approaches and sought a new framework that would transcend the old oppositions. Many of these biologists were sympathetic to mechanical explanations, as far as they could go, but they doubted that all questions in biology could be reduced to mechanical explanations.[51] In 1891 Hans Driesch published results of experiments on the eggs of sea urchins, which seemed to contradict Roux's mechanical interpretation of embryological development. Driesch separated two cells of a sea urchin embryo from each other and found that both cells developed into normal larva. This discovery led him to conclude that some "harmonious equipotential system" governed the development of embryos and allowed them to respond to changing external conditions.[52] Driesch eventually abandoned attempts at mechanical explanations of his harmonious equipotential system and turned to a belief in a neovitalism that was rejected by most biologists. But Driesch's emphasis on holism and organization represented a new direction in biology.

One of the best examples of the new interest in holism and organization can be found in the work of the American physician and physiologist Walter B. Cannon.[53] During his work as an Army physician in World War I, Cannon became interested in the problem of shock, and began to study the role of the sympathetic nervous system and the endocrine system in regulating organic functions such as temperature, breathing, heart rate, and metabolism. Cutting parts of the sympathetic nervous system in animals, he found that the animals' physiological functions were unable to adapt to changing environments. If such an animal were placed in a hot or cold environment, it would not be able to maintain a constant bodily temperature. This research led Cannon to develop the idea that there was a self-regulating process, or homeostatis, that allowed an organism to maintain a fixed internal system. This fixed condition was not established by isolating the organism from external changes but through an interactive process between internal and external conditions so that the rates of physiological processes changed in such a way that a condition of internal equilibrium could be maintained. Although Cannon believed that homeostatis functioned according to the laws of physics and chemistry, he did not believe that it could be reduced to the actions of any specific molecules. Neither vitalistic nor mechanistic, Cannon's work focused attention on the role of organization in biology.

By the 1920s and 1930s, the concept of organization was coming to play a significant role in biology. In 1924 the German embryologist Hans Spemann conducted a series of experiments, assisted by his graduate student Hilde Mangold, in which a section of a newt embryo, known as the dorsal lip, was transplanted into a different area of another embryo.[54] They discovered that the transplanted material "induced" the development of an entire second embryo attached to the original. Spemann argued that the dorsal lip contained some material which "organized" the second embryo's development. Since the "organizer theory" seemed to explain how cells became differentiated into specific tissues and organs, it created a great deal of excitement in biology, gaining the support of scientists such as Julian Huxley, and winning a Noble Prize for Spemann.[55] At first many people assumed that the organizer could be explained in terms of some chemical mechanism, but Spemann and others viewed the organizer as part of an interactive process in which the tissue being induced to change had a reactive effect on the organizer, transforming its effect on other tissues. As such, Spemann rejected the idea that the organizer could be reduced to some physical-chemical mechanism.

The concept of the organizer as part of a holistic system led some biologists such as Paul Weiss to postulate the existence of a morphogenetic field which governed embryological development.[56] Weiss, who had training in both engineering and biology, proposed his influential field theory in his book *Morphodynamik* (1926). In a series of experiments, tissue that was the source of the development of a tail in an amphibian embryo was transplanted to the area that normally developed into a limb; Weiss observed that a limb would result if the tail material was transplanted early in its development, but a tail would result if the material had been allowed to develop for a longer period of time. This observation led Weiss to postulate the existence of a field that organized undifferentiated cells into a specific pattern. According to this model, embryological development was a complex process that involved both the interaction between material parts of an organism with each other, and the interaction of the field with the whole organism. Although fields arose from the material nature of organisms, they required holistic explanations rather than reductionist ones. For example, the division of a field into two parts resulted in two complete fields, not two halves of one field, just as the division of a magnet would result in two smaller but whole magnets. Like the concept of the organizer, the idea of a field fell outside both mechanistic and vitalistic explanations.

The transformation in biology during the late nineteenth and early twentieth centuries reflected the emergence of a new, transcending world view; historian Garland Allen uses the term *holistic materialism* to refer to this new philosophy, while many of the biologists themselves used the term *organicism*.[57] In 1903 the French scholar G. Delage argued that such nineteenth-century biologists as Claude Bernard no longer could be classified as vitalists or mechanists but should be labeled "organicists," and in 1917 the American physiologist Lawrence J. Henderson suggested that the work in twentieth-century biology, such as that of Cannon and Weiss, focused on the problem of organization.[58] In the Silliman lectures at Yale during that same year, English biologist J. S. Haldane argued that neither mechanism nor vitalism could adequately explain biological phenomena, and he called for a new organicist approach.[59] Not long after, the American zoologist W. E. Ritter introduced the term *organismalism* to describe a holistic approach to organism, and in 1928 Ludwig von Bertalanffy put forward the idea that the "organismic conception" could transcend mechanism and vitalism.[60]

Holistic materialists or organicists emphasized the relationship and interactions between parts of an organism and how those parts are organized into a whole. Like mechanists, they rejected the idea of some mysterious vital force, but like the vitalists they rejected the idea that an organism can be completely understood by reducing it to its simplest parts. What is important to organicists is the interaction between the parts. Such interactions lead to new phenomena that cannot be understood by analyzing the parts in isolation from each other. A part of an organism that is interacting with other parts is fundamentally and qualitatively different from a part isolated from the whole. Organicists would not reject the idea of studying parts in isolation, but they would argue that such an analysis could only lead to a partial description and understanding of an organism. For the organicist, an analysis of the individual parts of an organism might be necessary but was not sufficient for a complete understanding. By emphasizing how the interactions of material parts can lead to the emergence of vital behavior, this model of biology contributed to a new world view that was in one sense both vitalistic and mechanistic, and in another sense neither vitalistic nor mechanistic.

The Wave Theory of Light

At the same time that the theory of organisms was undergoing a change, a corresponding transformation took place in the theory of the inorganic physical world. At the beginning of the nineteenth century, theories of the physical world were dominated by a reductive and mechanical philosophy. It was assumed by most scientists that physical phenomena could be explained in terms of motion, units of matter (atoms) and Newtonian forces acting either by attraction or repulsion between the atoms. Eighteenth-century scientists had had difficulty explaining certain phenomena such as heat, electricity, and magnetism, in terms of mechanical philosophy and were forced to postulate the existence of "imponderable fluids," such as the caloric or the aether, to explain these phenomena.[61] Most scientists assumed that even these "fluids" consisted of extremely tenuous mutually repulsive particles that ultimately acted in terms of mechanical laws.

During the second half of the nineteenth century, this picture of mechanical philosophy began to change. More and more physical phenomena became divided into two areas—one involving a theory of matter that could be explained mechanically in terms of discrete atoms, and the other involving a nonmechanical theory of radiation and light that could be explained in terms of continuous waves. By the end of the century, the desire to overcome the distinction between mechanical particles and waves of energy led to a new model of the physical world which synthesized the concepts of particles and waves.

As we have seen, mechanical philosophy had become well established during the nineteenth century through the successes of the kinetic theory of heat and Dalton's chemical atomism. But, just when mechanical philosophy seemed to be most firmly demonstrated, challenges to it were arising in other areas of the physical sciences. One such challenge involved theories of light. Most eighteenth-century scientists had assumed that light was a series of particles which could be described by mechanical philosophy. The particle theory of light seemed to explain the apparent straight line motion of light rays. If different sized particles were associated with different colors, the model also explained how a prism could break a beam of white light into a rainbow. But there were several optical phenomena that required ingenious, if not tortuous, explanations if the particle theory of light were true. Even at the time of Newton, Christiaan Huygens had argued that double refraction (that is, the ability of some crystals to split a light beam into two separate rays), could be more easily explained if light was thought of as a wave or a vibration.

It was not until the beginning of the nineteenth century that a wave theory of light began to be taken seriously.[62] Almost simultaneously three scientists, Thomas Young, Augustin Fresnel, and Francois Arago, proposed a wave theory of light, based on experiments regarding the phenomena of interference and diffraction. If two plates of glass were separated by a thin wedge of air a pattern of concentric circles, known as Newton's rings, could be observed. It was very difficult to explain this phenomenon with the corpuscular theory of light, but Young noted that it could easily be explained if one assumed that waves of light interfered with each other at certain points so that the peak of one wave would be at the same place as the trough of another wave. Soon after, Fresnel applied a similar theory to the problem of diffraction (the bending of light around the edge of an object). An implication of his theory was that at a certain distance a bright spot of light should appear in the center of a shadow cast by a circular disk.[63] A corpuscular theory of light could not explain the existence of such a spot, and Fresnel demonstrated that the spot did occur.

But the acceptance of a wave theory of light led to some significant problems. The most serious was the question: What material were the waves vibrating? Scientists had earlier noted that sound was the vibration of air. A ringing bell in a closed bell jar would cease to make sounds if the air was evacuated. However, light would continue to pass through the apparent vacuum. Some scientists postulated that the universe was filled with a luminiferous aether, an extremely tenuous material that would support the vibration of light.[64] But such a material would have to be rigid, like a solid, in order to support vibrations, and yet subtle enough to allow the planets and other bodies to move through it with no resistance. It was hard to conceive of a material with the required mechanical properties.

Field Theories

The solution to the problem came from another area of physics. At the same time that scientists were studying the wave theory of light, interest in electricity and magnetism revived. Ever since the time of Descartes the actions of electricity and magnetism, especially the ability of magnets or electric charges to both attract and repel each other across a distance or through other materials, had posed problems for mechanical philosophy. But in the nineteenth century the two phenomena were found to be related to each other—in 1820 Hans Christian Oersted observed that an electric current could produce a magnetic effect—and a new model for explaining electricity and magnetism began to emerge.

One of the people most responsible for the new model of electricity and magnetism was Michael Faraday, director of the Royal Institution in London, who in 1831 began a series of systematic experiments to find the inverse of Oersted's discovery—the creation of an electric current by a magnet.[65] That same year he was able to show that a coil moving in the vicinity of a magnet would have an induced electric current. Although Faraday's discovery of electromagnetic induction led to important practical applications such as the dynamo, his model for explaining the phenomenon had a profound effect on the theory of electricity and magnetism. The discovery of electromagnetic induction again raised the problem of action at a distance; the magnet was inducing an electric current not through direct contact with the coil but simply by being near to it. As an explanation, Faraday advanced the idea that electricity and magnetism were associated with ''lines of force'' which in the case of a magnet could be made visible if iron fillings were placed on a piece of paper covering the magnet. By the 1840s and 1850s, he began to think of these lines of force as having a real physical existence. It was through these physical lines of force which filled space that an object, like a magnet, could induce a current into a wire coil at some other place.

This model of electricity and magnetism based on lines of force, or fields as they would become known, raised some serious questions about the mechanical idea of discrete atoms moving in an empty space. In 1844 Faraday noted that the classical theory of atomism produced contradictions when called on to explain the nature of electrical conductors and insulators.[66] Studies of the density of materials indicated that the atoms in solid matter could not be in direct contact with each other but must be separated by empty space. If a material was a conductor of electricity, then the space between the atoms must also conduct electricity; if the material was an insulator, then the space between the atoms must also be a nonconductor of electricity. This reasoning led Faraday to reject the mechanical idea of atoms in a void and to replace it with the notion of atoms as centers of force. The forces associated with these point-centers could have localized areas of repulsive force and act like solid matter, but the forces would also extend beyond the space associated with material atoms and fill all space. This space-filling aspect of force could then be associated with gravitation, electricity, magnetism, and, most importantly, light.

Using this model, Faraday extended his researches on the interactions between electricity, magnetism, and matter. Since the idea of space-filling forces helped to explain the relationship between electricity and magnetism, there might be other relationships to discover between the forces. In 1845 Faraday discovered that polarized light traveling through a magnetic material would have its plane of polarization changed. The next year he raised the possibility that light might be a vibration of the lines of force associated with electricity and magnetism.[67]

Faraday's suggestions concerning lines of force and the nature of light were expanded by James Clerk Maxwell into the concept of electric and magnetic fields.[68] In his early researches, Maxwell tried to find a mechanical model for his theory of fields. He postulated that Faraday's lines of force were tubes of rotating aether whose lengthwise contractions and sideways expansions could explain attraction and repulsion. Since adjacent tubes would be rotating in the same direction he had to postulate a set of "idle wheels" or ball bearings between them.[69] It became difficult for anyone, even the strongest supporters of mechanical philosophy, to believe that the universe was filled with a material composed of rotating gears and idle wheels. But the model did provide Maxwell with the beginning point for a monumental mathematical analysis of the actions of electric and magnetic fields. The result of this analysis was a series of papers, including "A Dynamical Theory of the Electromagnetic Field," in which he derived a set of four equations that have since been known as "Maxwell's equations." These explained the known phenomena of electricity and magnetism, and they had some important consequences. The equations were found to be wave equations and, more importantly, the waves they described were the same as those describing the wave theory of light. Maxwell claimed that "light consists in the *transverse undulations of the same medium* which is the cause of electric and magnetic phenomena."[70] Maxwell had unified under the actions of electromagnetic fields the phenomena of electricity, magnetism, and light.

An Electromagnetic View of Nature

The idea of electromagnetic fields helped to undermine traditional mechanical models of natural phenomena. Maxwell had hoped to explain electromagnetic fields in terms of some mechanical model, but in the end he was unable to realize his goal.[71] Maxwell never made it clear whether electricity was a substance, a form of energy, or some unknown physical quantity.[72] If electromagnetic fields and the luminiferous aether could not be explained in terms of mechanical philosophy, several physicists began to investigate the possibility that the mechanical actions of matter might be explained in terms of electromagnetic concepts.[73] As early as 1867, British physicist William Thomson, who would later become Lord Kelvin, published a paper "On Vortex Atoms," which suggested that atoms might be swirling filaments of the aether similar to smoke rings in the air. By the 1880s and 1890s, physicists such as Joseph Larmor attempted a synthesis of Thomson's vortex atoms and Maxwell's electromagnetic field theory.[74] Unlike ordinary matter, Larmor's aether was a pure continuum. This aether was not made up of matter; rather matter was a structure of the aether.[75] He argued that there could be centers of strain in the aether which were endowed with an electric charge. These centers of strain, which he labeled *electrons,* were both the fundamental units of matter and the source of the electromagnetic property of the aether.

By the end of the nineteenth century, Hendrik Antoon Lorentz had eliminated all mechanical properties from the aether and defined it purely in electromagnetic terms.[76] He assumed that electrons were electrically charged particles embedded in the aether, and were also contained in ordinary matter. These electrons generated electromagnetic fields in the aether and such fields could then act on ordinary matter through the electrons it contained. There was no mechanical connection between ordinary matter and the aether; rather the connection was purely electromagnetic. By 1895, Lorentz had rejected mechanical concepts as fundamental to nature and argued instead that electrodynamics could serve to unify physics.

The Theory of Relativity

By the beginning of the twentieth century, the mechanical world view was being challenged by an electrodynamic view of nature. The phenomena of the physical world seemed to fall into two categories: those that were essentially mechanical and could be explained in terms of discrete pieces of matter moving in space according to Newton's laws; and those that were essentially electrodynamic and could be described in terms of electromagnetic fields obeying Maxwell's equations. Studies concerning the motion of bodies and light through the aether appeared to indicate that these two views of physical phenomena were incompatible. According to the laws of mechanics, an object in space, moving in the same direction as the Earth, would appear—when measured from the Earth—to be moving slower than its true speed, while a body moving in the opposite direction would appear to be moving faster. But Maxwell's equations implied that an observer moving toward or away from a beam of light would always measure its speed as a constant. Several experiments, including a famous one conducted by Albert A. Michelson and Edward W. Morley in 1887, demonstrated that the motion of the Earth had no effect on measurements of the speed of light.

The contradictions between the laws of mechanics and the laws of electricity and magnetism were resolved by Albert Einstein in his paper of 1905 "On the Electrodynamics of Moving Bodies," in which he stated his theory of relativity.[77] Since the age of fifteen, Einstein had been thinking about the differences between the motion of mechanical objects and the motion of light. In trying to decide what would happen if he were moving at the speed of light and looked in a mirror, he concluded that he would not see anything since the light from his face could never reach the mirror. According to Newtonian mechanics, a body, given enough energy, should be able to attain any finite speed, but moving at or faster than the speed of light leads to absurd consequences.

To resolve this conflict between the laws governing the motion of light and the laws governing the motion of matter, Einstein was forced to transform both the mechanical world view and the electromagnetic world view, and to reformulate physics. Einstein began by rejecting the concept of either a mechanical or electrodynamic aether. Light simply moved through empty space. But he accepted as fundamental two principles—one drawn from Newtonian mechanics and the other drawn from Maxwell's electrodynamics. The first was a principle of relativity, which stated that the laws of physics should be the same in any frame of reference moving relative to another frame at a constant velocity. This principle was based on the well-known fact that Newton's laws of mechanics would hold equally well for a person stationary on the ground as for a person on a train moving at a constant velocity. But Einstein argued this principle applied to all physical laws, including the laws of electrodynamics. His second principle was that the speed of light was a constant, which had been observed by experiments, and could also be derived from Maxwell's equations.

These two principles could not be reconciled within the traditional mechanical world view. If the speed of light is constant and the relativity principle applies to light, then the speed of light would be the same even if one were at rest or were moving parallel to it at some great speed. This violated Newton's laws of motion, not to mention common sense. In order for both principles to be true, Einstein had to modify the fundamental concepts of length, time, and mass that were the foundations of the mechanical world view. He concluded that these concepts were not absolute, but depended on the motion of one frame of reference relative to another. According to Einstein, if a ruler, clock, and mass were moving relative to other rulers, clocks, and masses, the moving ruler would be shorter, the clock slower, and the mass heavier than those that were considered at rest. Such changes would explain why the speed of light was always measured the same, even by a moving observer, since that observer would be using a shorter ruler and a slower clock to calculate the speed of light.

These changes in length, time, and mass were not simple "illusions"; rather they reflected the fact that in Einstein's theory such fundamental units could no longer be considered as absolute and unchanging. Even mass, a foundation of the mechanical world view, was not absolute. A mass moving relative to another increased, which led Einstein to conclude, later in 1905, that energy could be transformed into mass and vice versa ($E = mc^2$). With the formulation of Einstein's theory of relativity and the later experiments that seemed to confirm it, the mechanical world of objects defined by absolute lengths, times, and masses was being replaced by a changing world defined by events and relationships.

Radioactive Atoms and Quantized Radiation

As Einstein was developing his theory of relativity, an even more radical break with the mechanical world view was emerging. Since the time of Newton, the idea that physical phenomena could be explained in terms of basic units, or atoms of matter, had been one of the foundations of mechanical philosophy, but, by the 1890s, scientists began to doubt that atoms were the immutable, eternal elements of matter. Some, such as Wilhelm Ostwald and Ernst Mach, even questioned whether atoms existed.[78] Since no one at the time could point to any direct experimental evidence for the existence of atoms, these scientists argued that physical theories should reject the notion of the atom and focus on what was an observable phenomenon—the transformation of energy from one form into another. This energeticist point of view was a minority position, soon made weaker by discoveries such as Brownian motion and the measurement of a fixed unit of electric charge, which supported the mechanical notion of the atom. But as Einstein noted, the works of people like Mach made him question the "dogmatic faith" of physicists in mechanics as the foundation of science.[79]

For the majority who continued to accept the existence of atoms, the actual nature of atomism was undergoing change. Since the time of the Greeks it had been assumed that atoms were the smallest units of matter, indivisible, immutable, and eternal. But in the 1890s experimental discoveries began to challenge this view. Scientists discovered that when electricity was applied to an evacuated tube containing two metal plates, some type of "rays" passed from one plate (the cathode) to the other. In 1897 J. J. Thomson at Cambridge proved that these cathode rays were, in fact, a stream of particles (later called electrons) whose weight was significantly less than the hydrogen atom, the lightest atom. Another question was raised when a group of scientists, including W. Roentgen, Henri Becquerel, and Pierre and Marie Curie, discovered that certain elements, such as uranium and radium, emitted forms of radiation that changed an atom of one chemical substance into an atom of another chemical substance. The discovery of radioactivity implied that the atom could no longer be considered immutable and eternal.

At the same time that questions were being raised about the mechanical, billiard ball model of matter, scientists were also discovering problems with the notion of continuous fields. The difficulty centered on describing the nature of "blackbody" radiation.[80] If, as an ideal source of radiation, a blackbody which emitted and absorbed radiation at perfect efficiency, was heated to a given temperature, the wave theory of light predicted that all possible modes of vibrations should be given off, and there could be an infinite number of possible modes of vibration corresponding to waves of shorter and shorter wavelengths. Therefore, according to theory, a blackbody should radiate an infinite amount of energy, mostly in short wavelengths in the ultraviolet part of the spectrum. However, experiments had shown that at any given temperature, there was a maximum wavelength, well below the ultraviolet region, which blackbodies emitted.

The solution to this "ultraviolet catastrophe" was suggested by Max Planck in 1899 when he argued that radiation was not emitted and absorbed continuously but in discrete packets of energy or quanta. There was a lower limit on the energy that could be emitted or absorbed by a blackbody and therefore energy would not "flood" into the ultraviolet end of the spectrum. A few years later in 1905, Albert Einstein extended Planck's notion and argued not only that radiant energy was emitted and absorbed in discrete packets, but that it existed as such packets or "photons." That is, electromagnetic radiation, or light waves, had the peculiar property of being waves yet traveling in discrete packets each with its own energy and momentum and therefore also acting like a stream of particles. Thus, at the same time that the mechanical theory of matter was shown to be something more complex than billiard ball atoms, the electromagnetic theory of light was shown to be more complex than simple waves.

Quantum Mechanics

It seemed the physical world could no longer be explained by either mechanical particles or continuous waves. A new model was needed, one which transcended the distinction between wave and particle. It arose in the second decade of the twentieth century in the new theory called quantum mechanics.[81] Planck and Einstein had already shown that radiation and light could have both wavelike and particlelike properties. In 1924 Louis Prince de Broglie suggested that matter might also have a similar dualistic nature—that is, a stream of particles moving through space might exhibit wavelike properties. The idea was confirmed three years later when C. J. Davisson and L. H. Germer showed that a beam of electrons reflected from the surface of a crystal produced a diffraction pattern similar to the pattern produced by reflecting X rays from the same crystal. De Broglie's theory helped to bring about a revolution in physics.

Between 1919 and 1925 Ernest Rutherford and his student Niels Bohr had developed a model of the atom based on an analogy with the solar system.[82] They argued that most of the weight of the atom was concentrated in a very small positively charged nucleus surrounded by lightweight orbiting electrons. The amount of positive charge (or number of protons) in the nucleus determined the chemical characteristic of the atom while the orbiting electrons provided the basis for chemical combinations with other atoms. But there remained problems with this planetary model. According to nineteenth-century physics, the orbiting electrons should have radiated energy and collapsed into the nucleus. Bohr had to introduce an arbitrary assumption that certain orbits allowed the electrons to remain stable. With the new discoveries of the dualistic properties of both radiation and matter, the basis was provided for a new model of the atom and, with it, all the rest of the physical world.

The work of Werner Heisenberg and Erwin Schrödinger between 1925 and 1935 brought an end to the mechanical model of the atom and put in its place a new quantum model centered on the concept of a wave-particle duality.[83] Using de Broglie's idea of matter-waves, Schrödinger was able to derive an equation that described the action of an electron in an atom. Rather than treat the electron as a discrete particle, Schrödinger's wave mechanics treated the electron as a standing wave. This explained why Bohr found that electrons could be in only certain orbits around the nucleus. Only orbits that corresponded to a whole number of wavelengths were acceptable, otherwise the standing electron wave would not "fit" into the orbit. Therefore, orbiting electrons could not absorb or emit just any amount of energy and move to a slightly larger or smaller orbit. They had to make a "quantum jump" to the next acceptable orbit. Schrödinger's wave mechanics spelled the end of the simple mechanical planetary model of the atom. In Schrödinger's theory no one could say where the electron was in its orbit, since it was a standing wave existing at all parts of the orbit. Max Born reinterpreted Schrödinger's matter-waves and argued that the waves were related to the probability of finding the particle at any given point. In Einstein's words, "God was playing dice with the Universe."

The formulation of quantum mechanics raised some serious philosophical and conceptual issues: Were material objects particles or waves, and how were their movements determined? In 1927 Bohr and Heisenberg, who was working with Bohr in Copenhagen, discovered two principles which formed the basis of the most widely accepted view of quantum mechanics, known as the "Copenhagen interpretation."[84] Heisenberg, who had earlier formulated a version of quantum mechanics independent of Schrödinger, derived the "uncertainty principle" from the mathematics of his theory. In essence, the uncertainty principle stated that it was impossible to measure at the same time, with infinite accuracy, the position and momentum (velocity) of a particle. At the most fundamental level, the determination of the position of a particle, by seeing it or bouncing one photon of light from it, would disturb it. Thus, its original direction of motion would be uncertain. This principle was a severe attack on the mechanical world view, which assumed that the position and direction of all particles could be precisely known. In the quantum world uncertainty and therefore indeterminacy were facts of nature.

While Heisenberg was discovering his uncertainty principle, Bohr was formulating his principle of complementarity. To ask whether objects were waves or particles was, for Bohr, to pose the wrong question. As with the uncertainty principle, the experimental situation would have an effect on the findings. There were certain complementary concepts, such as waves and particles, each of which could be observed with the right experimental situation, but since they were complementary there was no experiment that could observe both at the same time. For example, if you look for the particle properties of an electron you can find them with a given experiment, and if you look for the wavelike properties of an electron you can find them with another experiment, but there is no experiment that will demonstrate both the particle and wave properties of an electron. According to the

Copenhagen interpretation, the objects of nature are described by a wave-particle duality. In one sense they are both particles and waves; in another sense they are neither particles nor waves. We may, in fact we must, talk about objects in terms that are mutually exclusive without logical contradiction.[85]

More recently physicist David Bohm has argued that the development of quantum mechanics and the theory of relativity has led to a new order of the world. He notes that relativity and quantum theory

> imply the need to look on the world as an *undivided whole,* in which all parts of the universe, including the observer and his instruments, merge and unite in one totality. In this totality, the atomistic form of insight is a simplification and an abstraction, valid only in some limited context.[86]

Bohm contrasts the new order in physics to the older mechanical order. The new "implicate order" is not understood in terms of the arrangement of independent objects throughout space or independent events over a period of time. Instead, the order of the whole system is thought to be implicit or enfolded in each region of time and space.[87] For Bohm, the new implicate order can be seen as analogous to a hologram. In a hologram a photographic plate, containing a pattern, is illuminated with a laser. The reflected light from the plate forms an interference pattern which produces a three dimensional image. But unlike a traditional photograph, there is no one-to-one relationship between any particular region of the photographic plate and the final image. All parts of the plate contribute to all parts of the image. If the plate is broken in half, the entire image will still be produced but not in as sharp detail.

Although controversial, Bohm's idea of the implicate order has attracted some significant supporters and has pointed toward a new model for understanding the world, one that transcends both mechanistic and vitalistic explanations.[88] Bohm himself notes that

> when understood through the implicate order, inanimate matter and living beings are seen to be, in certain key respects, basically similar to their modes of existence. . . . It may indeed be said that life is enfolded in the totality and that, even when it is not manifest, it is somehow "implicit" in what we generally call a situation in which there is no life.[89]

The Bionic World View

We have now traced how the mechanical and organic world views evolved and interacted with each other throughout history. In many cases discoveries, concepts, methodologies, and models that originated in one world view were incorporated into the other. By the beginning of the twentieth century, the distinctions between the organic and mechanical world views began to break down. The biological theories associated with holistic materialism and organicism could not be classified as either mechanistic or vitalistic. At almost the same time, the theory of relativity and quantum mechanics brought an end to the classical approach to physics and chemistry based on mechanical philosophy. Reductionism was called into question in both views. Objects in the quantum world could not be reduced to either particles or waves, and life could not be reduced to the study of isolated parts. Even if life could be reduced to physics and chemistry, those sciences were themselves questioning the reductionist approach.[90]

The transformation of both the mechanical and organic world views and the difficulty of maintaining a clear distinction between the two raises the possibility that we are witnessing the emergence of a completely new world view.[91] I want to argue that organicism, holistic materialism, relativity theory, quantum mechanics, and the implicate order represent elements of a new emerging world view—one that both incorporates and transcends the older mechanical and organic world views. I believe an appropriate label for this world view might be *bionic*. The term is appropriately ambiguous since it can refer to machines that substitute for organic processes as well as organisms that substitute for machines. In order to gain some insights into the characteristics of this bionic world view, we can examine the development of some recent philosophies that have tried explicitly to synthesize the mechanical and the organic.

One such new idea is the "process philosophy" introduced by Alfred North Whitehead during the 1920s. His Lowell Lectures at Harvard in 1925, which were published as *Science and the Modern World,* and his Gifford Lectures at the University of Edinburgh in 1927–28, published as *Process and Reality,* had a major influence on the development of organicism, holistic materialism, and quantum theory.[92] Whitehead based his philosophy on the belief that "science is taking on a new aspect which is neither purely physical nor purely biological."[93] He pointed to a new concept of the world when he said that "nature is a structure of evolving processes. The reality is in the process."[94] Unlike the static concepts that dominated philosophy for the previous several centuries, the heart of Whitehead's philosophy was the idea that "all things flow," which leads to the notion that the basic elements of the world were "pulsations" rather than some unchanging objects.[95] These "pulsations" could form units or events that had some degree of stability, even long-term stability. But unlike the immutable, eternal, and independent units on which mechanical philosophy was based, the "pulsations" were ultimately connected to all other levels of the entire process of which they were a part.[96]

Of particular importance to a bionic world view, Whitehead's philosophy was able to deal with concepts that other philosophies considered essentially contradictory. For Whitehead, the notion of inconsistency can be overcome by the idea of process. In normal usage, if two concepts or propositions are inconsistent they cannot both occur. But for Whitehead, "process is the way by which the universe escapes from the exclusions of inconsistence."[97] At a particular instant two ideas may be inconsistent, but "these two entities may be consistent when we embrace the whole period involved, one entity occurring during the earlier day, and the other during the later day."[98] Since process involves the notion of change or becoming, it must encompass apparently contradictory ideas; for example, the process of something coming into being must encompass the idea of that something not being and then being.[99] The contradiction exists only if we continue to insist on taking a particular static world view.

During the 1940s and 1950s many of Whitehead's ideas played a role in the development of another significant attempt to synthesize mechanical and organic ideas in what came to be called a *general systems theory*. As a philosophy, the idea of a general systems theory has been most forcefully put forward in the writings of Ludwig von Bertalanffy, who had been a significant figure in the development of an organismic biology during the 1920s.[100] His work in biology led him to question both the reductionism of the mechanistic approach and the mysticism of the vitalistic approach. Instead Bertalanffy focused on the role of organization. In the period after World War II, he began to see that problems such as order, organization, wholeness, and teleology could be found in a wide range of other areas including thermodynamics, engineering, psychology, sociology, and economics. This insight led Bertalanffy to conclude that various phenomena—from the actions of atoms to the actions of organisms, to the actions of machines, to the actions of human societies—could be better understood if they were treated as some kind of system. By 1954, Bertalanffy came together with Kenneth Boulding (an economist), Ralph Gerard (a physiologist), and A. Rapopart (a biomathematician) to found the Society for General Systems Research, and by the 1970s, some supporters such as Ervin Laszlo were calling for a systems philosophy as a new framework for contemporary thought.[101]

In the broadest terms a system can be defined as a *complex of interacting components,* but each of the components can also be a system or subsystem with its own interacting components.[102] Also, the interactions among or within components can be simple or complex, involving such things as feedback or backward propagation.[103] One important characteristic of a system is the fact that two components that appear to be the same will act quite differently in different systems or in different parts of the same system. It then becomes the aim of a general systems theory to formulate laws, principles, and models that govern the hierarchic structure, differentiation, stability, and goal-directedness of any type of system.

Like the idea of process, system provides a way of understanding the transformation of the organic and mechanical world views into a bionic world view. If the interactions within components of a subsystem or among some of the subsystems are weak, those parts of the system, as Herbert Simon has suggested, may be "nearly decomposable" into parts that are "approximately independent of the short-run behavior of the other components."[104] That is, in some situations and for some periods of time, parts of a system may function in a mechanical way while the other parts of the system function in a more organic fashion. But ultimately, such divisions are only approximate ways to describe the whole system.

In a bionic world view the mechanical and the organic would be related in terms of a system. This system could be called a *dualistic system* since at times some of the parts of the system might be nearly decomposable into either organic or mechanical subsystems. But it must be kept in mind that such divisions are essentially artificial and dependent on how one divides the system into subsystems. A different way of looking at the division of the system might cause some subsystems to change roles from mechanical to organic or from organic to mechanical. The organic and the mechanical are simply different manifestations of some other reality which is ultimately a bionic system and the process of interactions that form and maintain that system.

Elements of a bionic world view can be seen in two controversial theories that have recently emerged in physics and biology. Both the "anthropic cosmological principle" and the "Gaia hypothesis" have at their foundations the notion that the organic and physical worlds interact as a system.[105] Each theory questions the possibility of understanding either the physical world or the biological world independently.

According to the so-called weak version of the anthropic principle, the formation of intelligent life requires certain preconditions to exist within the universe. Since human beings do exist, we should not be surprised that certain physical and cosmological quantities display the limited range of values consistent with the conditions that allow life to emerge.[106] For example, this weak anthropic principle "explains" why the universe is ten to twenty billion years old. If, on the one hand, the universe were much younger, there would not have been time for the stars to have converted hydrogen and helium into carbon and oxygen which are the basis of organic life. On the other hand, if the universe were much older, most of the stars, like the Sun, which have planets that could support life, would have burned themselves out and died. Physicist W. Stephen Hawking has used the anthropic principle to explain why, in our universe, the initial velocity given to matter during the big bang is equal to the velocity matter needs to overcome its own gravitational attraction.[107] If the two were not equal, galaxies would not form since matter would be either homogeneously distributed throughout the universe or would collapse before galaxies could form. Since the formation of galaxies seems to be a precondition for the emergence of intelligent life, we should expect that out of all the possible relationships that could exist between the two velocities in our universe, they are equal. Most people would accept the weak version, but a more controversial strong version has been suggested.[108] This version argues that our very presence "causes" the universe to take on certain properties. The universe requires some kind of observer to bring it into existence.

In either version, the physical and biological worlds are treated as interacting components of a larger system. As astronomer Robert H. Dicke noted in 1957, cosmological quantities that describe the universe "were not random but conditioned by biological factors."[109] As John Barrow and Frank Tipler put it in *The Anthropic Cosmological Principle,* even the weak version

> deepens our scientific understanding of the link between the inorganic and organic worlds and reveals an intimate connection between the large and small-scale structure of the Universe. It enables us to elucidate the interconnections that exist between the laws and structures of Nature to gain new insight into the chain of universal properties required to permit life.[110]

The Gaia hypothesis, originally formulated in 1974 by British chemist James Lovelock and named by novelist William Golding after the Greek goddess of the

Earth, argues that the Earth's surface, atmosphere, and forms of life constitute a self-regulating system. Instead of viewing life as simply adapting to an independent physical environment, the Gaia hypothesis postulates that organic activity influences and in fact creates an environment that is conducive to some kind of life. For example, Lovelock has shown it is difficult, if not impossible, to explain the Earth's present atmosphere simply in terms of chemical activity. One would expect over time the free oxygen in the atmosphere to react with other chemicals, particularly nitrogen, and form poisonous nitrogen oxides. Nearby planets such as Mars and Venus, which do not contain life, have chemically stable atmospheres of ninety-five percent carbon dioxide, yet on the Earth this gas makes up only a small percentage of the atmosphere. According to the Gaia hypothesis, life, particularly widespread microbes which have existed for millions of years, has created and maintained the present atmosphere by means of photosynthesis, which removes carbon dioxide and produces free oxygen. Other organisms produce methane which seems to regulate the amount of free oxygen in the atmosphere. Not only is free oxygen maintained by a variety of living organisms, it is maintained within a certain narrow range in which life is possible—a little less oxygen and most animal life would die, a little more oxygen and forest fires would burn continuously even in rain forests.[111]

Lovelock and his former student Andrew Watson have shown that such forms of regulation do not require the existence of some mysterious force or intelligence, but can be explained in terms of simple systems models.[112] In an idealized "Daisy World" covered with white and dark daisies, an increase above the optimum growing temperature will lead to an increase in white daisies since they reflect light and lose heat. But an increase in white daisies will also lead to an overall decrease in atmospheric temperature since they reflect the Sun's rays. Conversely, a decrease below the optimum growing temperature will lead to an increase in black daisies since they absorb light and gain heat. But an increase in black daisies will lead to an overall increase in atmospheric temperature since they absorb the Sun's rays. As illustrated, a simple interaction between organisms and the physical environment can lead to a regulation of temperature in such a way that is conducive to the continuation of life. Like the anthropic principle, the Gaia hypothesis treats the physical and organic worlds as components of an interacting system. According to biologist Lynn Margulis and son Dorion Sagan, leading advocates of the hypothesis, "the Gaian view leads almost precipitously to a change in philosophical perspective. As just one example, human artifacts, such as machines, pollution, and even works of art, are no longer seen as separate from the feedback process of nature."[113]

Both the anthropic principle and the Gaia hypothesis reflect a new bionic world view. Much of the controversy over both theories arises from attempts to understand them in terms of the mechanical or organic world views. These theories cannot be understood in these terms. They only make sense in terms of a new world view which is, on the one hand, *both* organic and mechanical, and on the other hand, *neither* organic nor mechanical.

The Vital Machine

In a bionic world view, phenomena cannot be understood in terms of the symbols that have been used to represent the mechanical or organic world views. A new language and new concepts are required. Neither the mechanical clock nor an organism is an appropriate symbol for a world based on organization, implicate order, process, and system theory. We need a new symbol that can reflect the fundamentally dualistic nature of the bionic world.

I propose that we create a new concept—the *vital machine*—as a way to symbolize the bionic world view. The vital machine represents a dualistic system in which components or subsystems at one time thought to be organic or mechanical lose their individual identities and become part of a new category of phenomena. Unlike former views in which either organic phenomena or mechanical phenomena had to be subsumed under the other in a single vision of reality, in the bionic world the tension that exists between elements of the dualism cannot be reduced to one of its components. The idea of the vital machine is inconsistent or self-contradictory only if we insist on holding onto either the mechanical or organic world view. If we accept the historical evidence that both of these views are now spent, their incompatibilities are no longer relevant to the developments of the modern world, and trying to retain those views only hinders our abilities to face those developments. But the idea of the vital machine may contribute to a new model for understanding the relationship between technology and life as we move into the twenty-first century.

7

The Vital Machine

Man has, as it were, become a kind of prosthetic God. When he puts on all his auxiliary organs he is truly magnificent; but those organs have not grown on to him and they still give him much trouble at times.

SIGMUND FREUD, *Civilization and Its Discontents*

The concept of the vital machine provides a model for understanding the revolutionary developments that are emerging in the twentieth century from the new relationship between technology and organic life that has been established by the bionic world view. As historian and psychologist Bruce Mazlish has noted, throughout history we have seen the elimination of a series of discontinuities.[1] The result in each case has been to change drastically the way in which we think about ourselves and the order of the world. First, Copernicus eliminated the discontinuity between the terrestrial world and the rest of the physical universe. Next, Darwin eliminated the discontinuity between human beings and the rest of the organic world. And most recently, Freud eliminated the discontinuity between the rational world of the ego and the irrational world of the unconscious. But, as Mazlish has argued, there is one discontinuity that faces us yet. This "fourth discontinuity" is between human beings and the machine.[2]

In the twentieth century, the bionic world view has begun to overcome this fourth discontinuity. The line distinguishing technology from biology becomes increasingly ill-defined. In the past, supporters of the mechanical world view would have recognized the mechanical aspects of organisms while supporters of the organic world view would have recognized the organic aspect of technology, but these world views were incompatible. People had to choose between the two world views; no one could recognize that technology was *both* mechanical and vital or that a living organism was also *both* vital and mechanical. With the breaking down of the mechanical and organic world views in the nineteenth century and their replacement by a bionic world view in the twentieth century, we can begin to recognize the dualistic quality of both technology and life.

During the twentieth century two main paths have led to the evolution of the vital machine. On the one hand, systems that were predominantly mechanical have

113

been transformed by assimilating elements of the organic, while on the other hand, systems that were essentially organic have been transformed by assimilating elements of the mechanical.

Systems Building

A prototype of the evolution of the vital machine can be found in the technological and managerial revolution of the early twentieth century that systematized production.[3] Although there had been earlier experiments in systematized production, the modern concept was based on two key developments—the technological idea of the moving assembly line and the organizational ideas that arose out of Scientific Management. Most descriptions of systematized production focus on the moving assembly line and the use of interchangeable parts, but new managerial techniques were also necessary elements in the development of modern production. More importantly, it was the interaction between the technology of the continuous flow assembly line and the human behavior of workers that led to new concepts about technology and the human being that are symbolized by the vital machine.

The modern systematized factory had roots going back into the eighteenth and nineteenth centuries.[4] At the end of the eighteenth century, Oliver Evans used a series of conveyor belts, screws, and buckets to automate a grist mill. Grain unloaded at the mill was lifted by conveyor to the top of the mill where it was carried down to the grinding stones and then to bags. The mill could grind three hundred bushels in one hour. But in an age still dominated by a mechanical world view many people misunderstood the nature of Evans's invention. At a court case over the patent that Evans obtained on the mill, Thomas Jefferson testified that all of the elements, such as elevators and conveyors, had been previously invented.[5] By focusing on the individual mechanical elements of the mill, Jefferson and others failed to understand that Evans's essential invention was the combination of previously invented machines into an organized system that could perform a predetermined function through self-action.

During the early nineteenth century there were isolated attempts at improving factory production by establishing a systematic flow of the product from one machine to another.[6] But an important step toward integrating humans and machines arose in the 1860s in the meat packing industry. Speed was important in order to reduce spoilage, yet attempts to mechanize the meat packing process had encountered several problems. Since animals such as hogs or cattle had individual and irregular shapes, it was extremely difficult to design a cutting machine that could automatically adapt itself to an individual carcass; most of the actual butchering would have to be done by hand. But the process could be made more efficient if the carcasses could be moved more easily from one position to another and if the time between the parts of the butchering process could be reduced. By 1869, in the meat packing houses of Cincinnati, overhead rail systems were installed to move the carcasses from one station to another where they were split, disembowelled, and inspected. Although the process might be described more accurately as disassembly, historian Siegfried Giedion noted, "Here was the birth of the modern assembly line."[7]

The elements of the assembly line, the conveyors, and overhead rails had been used in earlier periods, but in the meat packing houses they were put to use in a new context. The moving assembly line allowed a process such as butchering, which was inherently nonmechanical, to be organized into a technological system. The human and the mechanical were combined into a symbiotic relationship with the assembly line. The actual butchering of irregular carcasses still required human actions which could not be duplicated by machines, but the sequence of steps between those individual human actions could be mechanized.

In many cases, such as meat packing, human action was required in the production process. During the last decades of the nineteenth century, attempts were made to systematize human activity in order to make it more efficient. The leader in this attempt to create a Scientific Management was Frederick Winslow Taylor.[8] In 1880 as a foreman at Midvale Steel, Taylor, applying a technique used by his high school teacher, began to use a stop watch to determine how long workers took to perform a specific task. Continuing these studies at the Bethlehem Steel Works, Taylor realized that workers were doing their tasks in a highly inefficient way. To improve worker efficiency, he began applying to human beings the scientific principles that were used to design machines.

Taylor's *Principles of Scientific Management,* published in 1911, was based on job analysis and time studies.[9] By observing a series of average workers, a supervisor could break down their job into a set of elementary motions. After each of these individual motions was timed with a stop watch, the useless movements could be eliminated and the quickest and best movements could be brought together. The result was the most efficient method of doing the task. Taylor's time and motion studies were further refined by Frank and Lillian Gilbreth, who placed emphasis on the way work was done rather than on the time it took to do a job. After studying the motions of a bricklayer, Frank Gilbreth was able to position the bricklayer, the mortar box, and the pile of bricks in such a way that all unnecessary movements were eliminated and the worker could reduce the number of move-ments per brick from eighteen to five.[10] This investigation led the Gilbreths into a study of pure motion.[11] The Gilbreths photographed workers who had small electric bulbs attached to their limbs while they did their jobs. On the photographs, the worker's movement produced a light track, or cyclograph, that could be analyzed for inefficient or extra motions—the smoother and simpler the curve, the more efficient the motion.

In their theories, Taylor and the Gilbreths made several assumptions about workers and the nature of human labor. Taylor assumed that workers would be motivated to work efficiently in order to gain increased wages. More importantly, they assumed that workers would do their work according to a set of predetermined instruction cards.[12] Scientific Management assumed that workers were similar to automatons in that they simply needed to be "programmed." In the end Scientific Management drastically transformed the concept of human labor. For the system to function, the individual worker could not act independent of the entire system. As Taylor himself noted:

The time is fast going by for the great personal or individual achievement of any one man standing alone and without the help of those around him. And the time is coming when all great things will be done by that type of cooperation in which each man performs the function for which he is best suited, each man preserves his own individuality and is supreme in his particular function, and each man at the same time loses none of his originality and proper personal initiative, and yet is controlled by and must work harmoniously with many other men.[13]

Although Taylor argued individuality still existed, his was a drastically transformed concept of the individual. He noted, "In the past the man has been first; in the future the system must be first."[14]

In Taylor's theory of Scientific Management, the system was made up of individuals—the managers, foremen, or other workers. By the beginning of the twentieth century, the theory of Scientific Management began to interact with the idea of the moving assembly line, and Taylor's concept of human control was taken over by the machine. The combination of the assembly line with Scientific Management brought about a new concept of production—one in which human beings and machines were integrated into a dualistic system. This new bionic form of production, a prototype of the vital machine, first arose in the manufacture of the automobile.

Henry Ford did not invent the automobile, but he did make it an item of mass consumption.[15] Ford had dreamed about creating the "car for the great multitude," but his major obstacle was finding a way of manufacturing something as complex as an automobile at a low cost. In 1913 a group of engineers at Ford's Highland Park plant in Detroit began experimenting with a moving assembly line for the production of magnetos. It had taken an individual worker eighteen minutes to assemble a magneto, but the assembly line reduced the time to five minutes. Within a year the idea of assembly line production was extended to the entire automobile. Ford himself noted that the methods of meat packing had suggested some of the elements of his assembly line production.[16] With the use of a continuous chain to pull the frame, a chassis could be completely assembled in one-and-a-half hours rather than the twelve hours needed for stationary assembly.

Ford's innovation went even further. The moving assembly line had been used in other industries, but what distinguished automobile production from these industries—like meat packing, that used some kind of continuous flow—was the complexity of the product being assembled. Each subassembly line for the engine and other components had to be synchronized and coordinated with the assembly of the chassis. For Ford's system to work, the workers' actions had to be systematized just as they were in Scientific Management. Although Ford never was directly influenced by Taylor, the ideas of Scientific Management were common knowledge at the time.[17] But a significant change had taken place. Instead of a foreman determining the actions of the workers, now it was the assembly line that controlled the pace of the work. Certain actions and motions could not be done effectively by machines, but the necessary human operations were determined by machines. While human beings were actually assembling the automobiles, their actions were "programmed" by the technology of the assembly line. With the success of Ford's Model T, his method of production quickly spread to other manufacturers and other industries.

Charlie Chaplin's movie *Modern Times* (1936) contains a telling portrait of this new system.[18] In the film, Chaplin's little tramp comes into contact with assembly line production, where his function is to spend eight hours a day tightening nuts with a wrench. He must perform the exact same motion hour after hour, day after day, always keeping pace with the assembly line. After a while the little tramp cannot be distinguished from the machines. His lunch is portioned into his mouth by a feeding machine while he continues his repetitive tasks on the line. By quitting time the little tramp has become a machine; he walks with a mechanical gait and continues to tighten anything that remotely resembles the nuts he has faced for eight hours. In the end, Chaplin's character rebels against the assembly line only to be swallowed up in the giant gears of the machine, literally becoming integrated into the machine.

Although still dominantly mechanistic, the new system of production introduced by Henry Ford had many of the characteristics of the vital machine. The system of production was neither purely mechanical nor purely human. In fact, when the system was functioning, the distinctions between the human workers and the mechanical machines tended to blur. On the one hand, the moving assembly line had a life of its own. There was an interdependence among the machines, and the entire assembly line had a degree of autonomy and self-activity that is usually associated with organic life. On the other hand, the workers lost many of their human characteristics and functioned as if they were machines. That is, the assembly line had taken on the characteristics of an organic machine while the workers were becoming mechanical organisms. In the end, this new system of production depended on the integration of organic machines and mechanical organisms, but in such an integration the boundary between the human and the machine began to disappear.

Although the new system of production began to erase the boundary between technology and the organic, the system was only a crude example of the vital machine. As demonstrated by Chaplin's *Modern Times,* it brought about the integration of the organic and technology more by transforming organisms into machines than by transforming machines into organisms. By not giving human intelligence to the machines, assembly line production stripped humans of their intelligence. Recent attempts have sought to bring about a more sophisticated version of a vital machine. One line of development has focused on the attempt to assimilate characteristics of human behavior, especially intelligence, into machines.

Computers

The modern idea of a machine that exhibits artificial intelligence arose out of an interaction between the development of the computer and new discoveries in the area of physiology and psychology. The computer is not wholly an invention of the twentieth century.[19] There is evidence that the Greeks and the Romans had designed mechanical devices using pegged counter wheels and display dials for use as mileage meters on both carriages and ships. The early mechanical clock developed out of a tradition of machines that were used to calculate the position of the planets in relation to the earth. By the seventeenth and eighteenth centuries, the development of mechanical computers was encouraged by the mechanical world view. If physiological processes could be explained mechanically, it seemed natural to assume that machines could be designed to carry out mental processes. Several famous mathematicians, including Blaise Pascal and G. W. Leibniz, designed mechanical calculating machines. Although these machines were quite simple, with functions usually limited to addition, subtraction, and sometimes multiplication and division, there was hope that machines would eventually imitate the higher aspects of human reason. Leibniz in his *De arte combinatorica* (1666) attempted to reduce human reason to a series of calculations.[20] In *Gulliver's Travels,* Jonathan Swift satirized the notion of creating a machine that would mechanically produce great ideas in philosophy, poetry, and law.

Before a practical computer could be developed, two independent sets of problems had to be solved and their solutions brought together. One problem concerned what we now call hardware. Most of the early machines were not designed by engineers, and they were subject to mechanical problems which limited their use. A practical computer would require the development of a new kind of machine that would be capable of processing information. The second problem concerned what we now call software. Most of the early machines were designed to accomplish one or two specific tasks, such as solving a particular equation. For different tasks to be accomplished, the machine had to be physically changed or redesigned. A truly general computer would have to be flexible enough to accomplish several tasks. This would require that the machine be able to use information in the form of a program that would direct the computer to do several different tasks.

During the nineteenth century, some advances were made in both the areas of hardware and software. In 1822 the English mathematician Charles Babbage wrote a paper for the Royal Astronomical Society arguing that a machine could be built to do the tedious tasks of calculating mathematical tables for astronomy.[21] The government supported Babbage's attempt to build a "Difference Engine," but he encountered many engineering difficulties and the machine was never completed. While he was still struggling with his Difference Engine, Babbage came up with a scheme for a much more sophisticated device—the Analytic Engine. The Difference Engine could do only one particular task, but the Analytic Engine was conceived to be a general computing device. Although this machine too was never completed, it contained the key elements of the modern computer. As conceived by Babbage, the Analytic Engine would have a store (what we call a memory) and a mill (a central processing unit). The engine would be "programmed" to do different tasks by using a series of punched cards similar to those widely used at the time to determine patterns to be woven on Jacquard looms. With the exception of Lord Byron's daughter, the Countess of Lovelace, few people took Babbage's work seriously, and not until the twentieth century would his idea of a computer be built.[22]

While Babbage was trying to create the hardware for the computer, a significant advance was made in the area of mathematical logic that would play a fundamental role in the development of computer software. In 1854 George Boole, a self-taught mathematician, published *An Investigation of the Laws of Thought, on which are Founded the Mathematical Theories of Logic and Probabilities.*[23] He believed that the "nature and constitution of the human mind" could be imitated by mathematical symbols.[24] In his work Boole was able to show that the concepts of logic could be expressed in a series of algebraic statements. With the use of these Boolean algebras, a computer could be designed to perform logical tasks and not be limited to just mathematical calculations. An important characteristic of Boolean algebra was that the logical statements could be expressed in a binary system of 1's and 0's which represented propositions that could be answered yes or no. Like Babbage's Analytic Engine, Boolean algebra had little practical influence on computer design until the twentieth century.

Elements of Babbage's Analytic Engine and Boole's logic were brought together at the end of the nineteenth century by Herman Hollerith, an engineer who invented a tabulating machine to process data from the 1890 U.S. Census.[25] Using holes punched in cards to indicate sex, race, age, and other characteristics, Hollerith's machine could tabulate and sort cards based on the data that had been punched into the cards. Also, by sorting the cards based on a series of binary yes-no answers, several kinds of data could be correlated. For example, if all the cards representing females were sorted according to race, and then the cards represent-ing whites were sorted according to age, the number of teenaged white females could be found. These machines were so successful that they formed the basis of the Tabulating Machine Company, the forerunner of IBM.

Although Hollerith's machines used punched cards and binary logic they were not general programmable computers. The modern computer developed in re-sponse to the needs of the military.[26] With new weapons and powerful guns introduced during World War I, the science of ballistics presented a significant problem for the military. Although it was easy to calculate the path of an ideal particle moving in a vacuum, the equations describing a real shell moving through the atmosphere were very complex and required long tedious calculations even for approximate solutions. By World War II, there were several projects under way to produce a computer to carry out scientific calculations required by the Army. Once the war began there also arose demands for computers to help crack enemy codes and to carry out shock wave calculations as part of the project to build an atomic bomb.

One of the first successes was achieved by a group at Harvard and IBM led by Howard Aiken. Combining many of the ideas of Babbage and Hollerith, they produced the Harvard-IBM Mark I in 1944. The Mark I brought to reality the ideas of Babbage but it had a severe limitation. Although it was electrically powered, it performed its calculations through a series of essentially mechanical telephone relay switches. Almost as soon as it was completed it had become technologically obsolete.

As early as 1937, John V. Atanasoff and his graduate student at Iowa State College began work on a machine that used vacuum tubes instead of electro–mechanical relay switches to conduct calculations.[27] But before he could perfect his computer, he was called by the government to conduct classified research for the war effort, and was never able to complete his work on the "Atanasoff Calculator."

Another wartime group, working at the University of Pennsylvania under J. Presper Eckert and John Mauchly, was developing the first truly electronic computer known as the ENIAC (Electronic Numerical Integrator and Computer).[28] Before moving to Pennsylvania, Mauchly had visited Ames, Iowa in 1941 to see Atanasoff's machine, and he became impressed with the idea of basing a calculator on electronic circuits instead of mechanical or electro–mechanical devices. Although it was not completed until after the war in 1946, the ENIAC revolutionized computer hardware. Because it used eighteen thousand radio switches, the ENIAC could make calculations two thousand times faster than the Mark I.

Although the ENIAC brought about great advances in the computer as a calculating machine, it could not yet be called a vital machine. As with all of the earlier machines, the ENIAC circuits had to be physically changed before it could accomplish a different task. Before the electronic computer would evolve as a vital machine, it would have to assimilate some characteristics of human behavior—namely, intelligence. The development of artificial intelligence required not only advances in computer hardware but the development of systems that would allow the computer, as a machine, to become integrated with the human organism, as an intelligent being. As computer hardware was being developed an equally important series of advances were taking place in the area of the philosophy and physiology of mind; these would lead to new theories of computer programming and control.

Beginning in the 1930s, several people involved with the development of the computer began to apply ideas drawn from research in the physiology and philosophy of mind to the design of computer systems.[29] Much of this work was based on two important studies of computer logic. One of the most important ideas was put forward by the British mathematician Alan Turing in his 1936 paper "On Computable Numbers." In it he described an ideal computer, what has become known as a Turing machine, that can move a tape containing symbols forward or backward one space at a time and either erase a symbol or print one on the tape. Turing showed that if a set of instructions could be written in a binary code, like Boolean algebra, this simple machine could be instructed to carry out all of the calculations that could be done on specially designed computers.[30] An implication of Turing's work, known as Church's thesis, was that a computer could accomplish any task that could be described by a precise set of instructions or algorithm. This thesis in turn implied that if human actions, including intelligence, could be described by some algorithm, they could be accomplished by a computer.

Turing's machine existed only in the abstract, but a year after his paper an important idea was put forward by Claude Shannon that would allow the Turing machine to become a reality.[31] In his master's thesis and in a paper "A Symbolic Analysis of Relay and Switching Circuits," Shannon demonstrated that the Boolean algebra that expressed logical statements in a binary system of yeses and noes could be used to describe the characteristics of electronic switches and circuits that could exist only in an off or on state. If electronic circuits could be explained by Boolean algebra, and Boolean algebra expressed laws of human thought, then Shannon's work implied that electronic circuits might be capable of expressing human thought.

The ideas of Turing and Shannon were combined in an attempt to design an improved version of the ENIAC.[32] Before it was completed, the military was already aware of limitations in the ENIAC design. The most significant problem was that the circuits had to be physically changed in order to reprogram the machine for a different task. In 1945 a group at the University of Pennsylvania began work on a new logical design for a computer, the EDVAC (Electronic Discrete Variable Calculator). One of the leading figures in this project was the mathematician John von Neumann.[33] In his *First Draft of a Report on the EDVAC,* von Neumann outlined the design for a general purpose computer, what has come to be called a von Neumann machine. An important new element of the EDVAC's design was its ability to store a set of instructions, or a program, along with the data.[34] With the binary codes of Boolean algebra, a program could be stored in the computer's memory in the same way as numerical data. The computer's central processing unit would be able to distinguish between data and instructions according to their location in the memory. In a von Neumann machine the central processing unit would read the first instruction in the memory; it would then retrieve the first set of data from the memory and execute the instruction; it would then store the results of this instruction in the memory and move on to the next instruction until the entire program had been carried out and the results stored in the memory. The concept of a stored program meant that the EDVAC could do a new set of tasks without its circuits being physically rewired. A single computer with a different set of programs could do the same tasks as a series of specially designed machines.

Although von Neumann would later reject the idea that a computer could surpass or even equal human intelligence, in his *Report on the EDVAC* he made an explicit connection between the computer and the human nervous system. Von Neumann referred to the elements of the computer as "organs" and said that the elements of the central processing unit along with the memory "correspond to the *associative* neurons of the human nervous system."[35] He also applied to the computer ideas from Warren McCulloch and Walter Pitts's research on the nervous system.[36] Their work showed that the actions of neurons could be explained in terms of a mathematical model that was similar to the propositional logic of information theory.

The ideas of McCulloch and Pitts played a significant role in eliminating the discontinuity between the machine and the organic. The most explicit statement of the new relationship between the computer and living systems was proposed by the mathematician Norbert Wiener in his 1948 book *Cybernetics: or Control and Communication in the Animal and the Machine.* This book developed out of a discussion between Wiener and Arturo Rosenblueth, a physiologist on the faculty of Harvard Medical School, who had worked with Walter B. Cannon.[37] The problem of the relationship between humans and machines had arisen while Wiener and Julian Bigelow, an engineer, were involved in a wartime project to improve the accuracy of antiaircraft guns. Since both the antiaircraft gun and the target plane had humans at the control, Wiener and Bigelow realized that it was "essential to know their characteristics, in order to incorporate them mathematically into the machines they control."[38] In studying the control of a machine, Wiener and Bigelow recognized the importance of the idea of feedback in which the actual performance of a machine is returned back as an input.[39] Their investigation of feedback in machines led them to speculate that it might also play an important role in the way in which the nervous system controls the human body. Along with Rosenblueth, Wiener and Bigelow were able to develop a model of the activities of the central nervous system "as circular processes, emerging from the nervous system into the muscles, and re-entering the nervous system through the sense organs."[40] This model led Wiener to conclude that the problems of communication and control were similar in machines and animals and that these problems centered not on issues of electrical engineering but on issues of information theory. He became "aware of the essential unity of the set of problems centering about communication, control, and statistical mechanics, whether in the machine or in living tissue."[41] Using a Greek term for *steersman*, Wiener gave the mane *cybernetics* to this new field of problems.

Artificial Intelligence

With the birth of cybernetics there arose the possibility that computers could imitate human intelligence, since elements of the computer were equivalent to the human nervous system. Wiener himself raised the possibility that machines might be able to learn.[42] By 1950 Alan Turing, who had discussed cybernetics with Wiener, began to speculate that machines could exhibit intelligent behavior. In an article entitled "Computing Machinery and Intelligence," Turing began with the statement, "I propose to consider the question, 'Can machines think?'"[43] To answer the question he created a game in which an interrogator could ask questions and receive answers only through a teletype. If a machine by its responses to questions could fool the interrogator into thinking it was human, then the machine would have passed "Turing's test" and would have to be considered as being intelligent.

Although Turing raised the argument that there was little functional difference between minds and machines, there was no computer at the time that could have come even close to being able to pass the Turing test. Some scientists like von Neumann doubted that computers would ever be able to exhibit true human intelligence but many of their students began to make serious attempts to design computer programs that would allow machines to exhibit truly intelligent behavior. In the summer of 1956 the Rockefeller Foundation sponsored a conference at Dartmouth College to study the problem of artificial intelligence.[44] The conference not only helped to coin the term *artificial intelligence*, but brought together the scientists, engineers, and physiologists, such as John McCarthy, Claude Shannon, Marvin Minsky, Allen Newell, and Herbert Simon, who would become the leading exponents of artificial intelligence. Although few concrete results came from the conference itself, it helped to set an agenda for future research in artificial intelligence.

The conference made people realize that artificial intelligence could manifest itself in widely differing circumstances. Newell and Simon developed a program called Logical Theorist that was able to prove theorems from Bertand Russell and Alfred North Whitehead's book *Principia Mathematica,* while Shannon and McCarthy worked on a program that would allow a computer to play chess against a human opponent. A significant conceptual breakthrough came about when Simon and Newell began to focus explicitly on the computer as a symbol manipulator rather than a mathematical calculator. This shift led many researchers to focus on the linguistic aspects of human intelligence and to attempt to design machines that could interact with humans through everyday language instead of specialized computer languages.[45] Work in this direction led to programs such as Joseph Weizenbaum's ELIZA and Terry Winograd's SHRDLU. ELIZA was one of the first programs to allow a computer to interact with a human using natural language. The program was designed to allow the computer to simulate the role of a Rogerian psychoanalyst and to carry on therapy sessions with a human acting as a patient. Weizenbaum himself admitted that ELIZA was somewhat superficial and only gave the illusion of understanding because a patient expects an analyst to respond in certain ways.[46] SHRDLU was a program that allowed a human to manipulate a series of colored building blocks, displayed on a television screen, through a robot that responded to English sentences. The computer would interact with the human. For example, if the robot was asked to move the cube and there was more than one cube, the computer would ask which cube should be moved. The program could also learn new terms; for instance, it could define a tower to be a stack of three blocks of the same color.

During the 1960s and 1970s, much of the research in the field focused on the attempt to incorporate intelligence into the computer by discovering the rules that govern human logic and designing software programs modeled on those rules. This so-called "rule-driven AI" took the rational behavior of the human mind as a model of intelligence that could be made part of a computer system.[47] Because of the enormous problems in discovering rules for the full range of human behavior, researchers focused on limited aspects of intelligence such as game-playing, theorem proofs, or limited bodies of knowledge. By the 1970s, the development of "expert systems" became quite successful. In 1974, a program named MYCIN was designed that could diagnose a wide range of diseases based on a list of symptoms. Other programs were designed to do chemical analysis or to recommend appropriate chemotherapy for cancer patients. Most of these expert systems were not intended to function independently but in conjunction with some human expert. As such they have led to situations in which knowledge and decisions emerge not from a human expert, or from the computer, but from the human and the computer functioning as a system.

More recently a second line of research in artificial intelligence has been gaining attention. This work attempts to incorporate intelligence into machines by modeling the hardware of the computer on the neural networks of the human brain.[48] In the 1940s McCulloch and Pitt's work on "neuro-logical networks" and Donald

Hebb's suggestion that neurons could learn by changing the strength of the connection between them when they were excited, inspired research into the possibility that technological equivalents of neurons might be designed which could be connected together in a way similar to the networks of neurons in the human brain.[49] During the 1950s and 1960s, this line of research was developed by Frank Rosenblatt, who designed a machine called a perceptron, which could be trained to recognize certain patterns.[50] He saw the perceptron as an analogue of the human brain and claimed "it would seem that the perceptron has established, beyond doubt, the feasibility and principle of non-human systems which may embody human cognitive functions."[51]

Rosenblatt's work formed the basis of what could be called emergent AI.[52] As opposed to rule-driven AI, which assumed that intelligence could be programmed into computers through a set of rules, emergent AI assumed that intelligence would exist in the strength of connections between processing units. As the computer is exposed to a series of inputs, the connections would be adjusted so that similar patterns of activation would result if similar inputs took place. In order to accomplish such tasks, a new computer architecture had to be introduced. Traditional computers had a single powerful processing unit that performed operations, one at a time, on data retrieved from a separate memory. In what has become known as parallel distributed processing, or connectionism, many small processors, each with its own memory, work simultaneously to solve a problem and information is then shared between processors.[53] As several experts have noted, connectionism, or parallel processing, represents a new holistic approach to artificial intelligence.[54] Rather than being stored in a specific location, information exists in the interactions between processors and therefore is distributed throughout the system.

During the late 1960s and 1970s the idea of connectionism, especially perceptrons, came under heavy criticism from Marvin L. Minsky and Seymour A. Papert, two pioneers in artificial intelligence.[55] Their mathematical analysis indicated that simple, so-called one-layer perceptrons were unable to carry out certain mathematical calculations, but by the 1980s, researchers in connectionism had moved on to more complex two-layer perceptrons that have overcome some of Minsky and Papert's limitations. Much of the early work was limited by a lack of hardware and only recently has W. Daniel Hillis, founder of Thinking Machine Corporation, built and marketed a parallel processing computer called the Connection Machine, with over sixty-five thousand processors.[56]

Although the artificial intelligence community still remains divided over connectionism, there have been some recent attempts to exploit a combination of rule-driven AI and emergent AI. Minsky has recently argued that the "advantages of distributed systems are not alternatives to the advantages of insulated systems; the two are complementary."[57] In an expanded edition of their critical book on perceptrons, Minsky and Papert raise the possibility that artificial intelligence may arise from a combination or linkage of parallel processing and serial processing. They explain that they "have come to try to develop 'society of mind' theories that will recognize and exploit the idea that brains are based on many different kinds of interacting mechanisms."[58] If a group of different types of neural nets were organized into a larger system so that some parts of the system were strongly connected to one another while other parts were relatively insulated, the advantage of both serial and parallel processing could be exploited. The idea of such a dualistic system seems to reflect a bionic world view in which intelligent behavior, traditionally seen as the exclusive province of the organic, emerges from a machine, transforming it into a vital machine.

The likelihood of computers' being able to achieve artificial intelligence has been the subject of intense debate. Those who follow what is known as the strong AI position believe that computers will achieve consciousness and be able to equal, or even surpass, human intelligence since they hold that the brain is nothing more than a complex computer, or in Minsky's term a "meat-machine," in which consciousness is simply the result of complicated computations.[59] Those who doubt the ability of machines to develop intelligence or consciousness argue that there are fundamental differences, in either hardware or software, between brains and computers. Some believe that computers and brains may function in similar ways but computers will never become fully conscious because they are not made out of the same organic materials that make up the brain.[60] Others believe that it is impossible to reduce all human experiences to some set of formal rules or algorithms similar to those that serve as the basis for computer software.[61] According to philosopher Hubert Dreyfus, either conclusion concerning the possibility of artificial intelligence would have a significant consequence.[62] If machines are able to exhibit intelligence, the implication is that human intelligence is also mechanistic. Conversely, if artificial intelligence is impossible, the implication is that the organic world of humans is the sole repository of intelligent behavior.

This type of debate over artificial intelligence reflects an attempt to see the computer in terms of the older mechanical and organic world views, but as I hope I have shown, the computer is not a product of either of these world views; rather, it is the result of a bionic world view. The most revolutionary aspect of the computer, more revolutionary than either of Dreyfus's conclusions, is its role as a vital machine. The most important aspect of artificial intelligence is not whether it will or will not equal human intelligence, but the fact that we can no longer distinguish where the human or intelligent aspects begin and where the mechanical aspects end. The computer has a dualistic nature that incorporates both the organic and the mechanical in such a way that the discontinuity between humans and machines begins to disappear.

An example of this new relationship between humans and machines was put forward by John Kemeny, a mathematician, philosopher, co-inventor of the computer language BASIC, and president of Dartmouth College. In his book *Man and the Computer,* Kemeny argues that since computers have the ability to think, remember, and communicate, they should be considered a new species.[63] He says, "I would like to argue that the traditional distinction between living and inanimate matter may be important to a biologist but is unimportant and possibly dangerously misleading for philosophical considerations."[64] But Kemeny's most important conclusion is that as a new species, the computer is establishing a symbiotic relationship with humans in which both species are contributing elements that benefit the other.[65] It is this symbiosis of humans and computers that I call the vital machine.

Biomedical Engineering

Attempts to assimilate human characteristics such as intelligence into machines offer one line of development leading to the emergence of the vital machine. But there is also another line of development. This path has focused on assimilating elements and characteristics of machines into biological systems and materials. The modern idea that parts of the human body can be replaced by some mechanical device arose out of an interaction between biology, medicine, and engineering. A few attempts at what is now called biomedical engineering were made before the twentieth century. Most of these early attempts involved the prosthetic replacement of a limb. There is some evidence that as early as 484 B.C. a Greek named Hegosistratus used a foot made of wood to replace one lost in the course of a prison escape.[66] Many prosthetic devices were developed out of interaction between surgeons and craftsmen who had some technical knowledge. In 1509 a famous iron hand containing gearing in the fingers was made for a knight named Goetz von Berlichingen.[67] A few years later the French surgeon Ambroise Paré worked with a locksmith to produce artificial legs with movable knee and foot joints.[68] He also produced a metal arm containing springs, levers, and ratchets to allow the elbow to bend and the fingers to open and close.[69] In these early prosthetic devices there was only a minimal integration between the mechanical device and the body. The devices were simply strapped to the body and their actions, such as the opening and closing of fingers, were accomplished by moving strings and levers with other parts of the body.

In the twentieth century the demand for prosthetic devices began to increase because of injuries sustained during World War I. As a result, there were attempts to design prosthetic devices that would function in a more natural fashion. A group of German military surgeons developed the technique of cineplasty, in which the prosthetic arm was connected to the remaining muscles in the stump by a metal pin, thus allowing the patient to move the device through muscular contractions.[70] With World War II and, later, the birth defects brought on by the drug Thalidomide, there were new attempts at developing self-powered limbs that could be directly controlled through the patient's own nerves. Many of these advances depended on combined work in physiology and control theory. In 1958 two Russian scientists, A. E. Kobrinsky and V. S. Gurfinkel, working at the Central Scientific Institute of Artificial Limbs and Rehabilitation in Moscow, had already begun working on an artificial hand that was activated by nerve impulses, so-called electromyographic signals, coming directly from the remaining muscles.[71] By the 1960s, Melvin J. Glimcher, a friend of Norbert Wiener, and several other researchers in this country began work on similar projects. These electromyographic devices created a much closer integration between the machine and the human. Although there were some limits in degrees of mobility, speed of movement, and sensitivity of touch, these devices were being controlled by messages from the brain.

During the first half of the twentieth century, research began on designing machines that could take over or replace the function of not only external limbs but also of internal organs. As early as 1913, John J. Abel, Leonard G. Rowntree, and B. B. Turner began animal experiments in which waste products in the blood were removed by passing the blood from an artery through a filter and returning it to a vein.[72] It was not until 1925 that G. Haas attempted the first hemodialysis of a human subject but the equipment was still quite crude. The first practical artificial kidney was built by Willem J. Kolff, a Dutch physician during the trying time of the Nazi occupation. Using cellophane tubing, obtained from sausage casings, as the filtering membrane, and a design based on the water pumps of automobiles, Kolff was able to treat a uremic patient. Although Kolff's dialysis machine could function as an artificial kidney, problems remained in connecting patients to the machine. Since a vein and an artery had to be surgically opened and then bound off after dialysis, the procedure could not be done on a routine basis. In 1960 a group, including engineer Wayne Quinton and physician Belding H. Scribner, developed a teflon shunt connecting a vein and an artery that could be surgically implanted in an arm. The shunt allowed the patient to be repeatedly connected to the machine so that the artificial kidney machine could be used as a long-term treatment for patients with chronic renal failure. In such cases the machine starts to become a necessary and integral part of the patient's life, even to the extent that some patients feel ''incorporated into the machine,'' or develop a strong ''attachment'' to it, and resist having a particular machine replaced with another.[73]

Another area of research on artificial organs has focused on the heart. As early as 1899 the physician H. A. Huntington suggested that a mechanical pump attached to a cadaver could imitate the actions of the heart so that medical students could practice surgical procedures under more realistic conditions.[74] In 1930, after a relative became ill with heart disease, aviator Charles Lindberg began research, with Nobel Prize-winning physician Alexis Carrel, at Rockefeller Institute on a perfusion pump that could keep organs alive outside of the body.[75] Although the news media labeled the invention an artificial heart, it was never intended to keep an entire patient alive. But the device was able to keep a pancreas alive and producing insulin for three weeks. By 1934, John H. Gibbon developed an artificial heart-lung machine in which blood could be taken out of the body and exposed to oxygen and then mechanically pumped back into the body.[76] The Gibbon's machine and the later Read-DeWall oxygenator were large machines; because of the damage they did to the blood cells by the action of the pump, they could be used only for short periods of time.

Other researchers focused on attempts at permanently replacing some parts or function of the heart. For years researchers had recognized that the heart's contractions were the result of electrical impulses originating in an area of the right atrium and that in some circumstances, such as a heart attack, these signals are blocked, resulting in a slow or erratic heart beat.[77] During the 1920s and 1930s, A. S. Hyman of Beth David Hospital in New York began studies on the artificial electrical stimulation of the heart. In 1933 he applied for a patent for an ''Artificial Pace Maker for the Heart.''[78] The device was quite large, weighing sixteen pounds, and could be used for only six minutes at a time, long enough to resuscitate a patient but not suitable to correct a chronic problem. Electrical stimulation of the heart was attempted only in isolated cases until 1952, when Paul Zoll of Harvard Medical school began a series of successful clinical applications using an external electrical pacemaker. Within a few years, Zoll, using disks placed on the skin, was able to use the external pacemaker for periods of up to seven days. With the development of the transistor, it became possible to miniaturize the pacemaker so that it could be permanently connected to the heart by implanting an electrode directly into the right ventricle. By 1959, battery technology had advanced to the point where Ake Senning and R. Elmquist were able to implant a rechargeable pacemaker into a human patient, and, by 1960, surgeon William Chardack and engineer Wilson Greatbatch designed an implantable battery-powered pacemaker that lasted more that two years and did not require recharging. By the mid-1960s, pacemakers had been developed that could be set at different rates or that would function only if the heart rate became abnormal.

At the same time that the pacemaker was being perfected, other researchers were developing artificial replacements for other parts of the heart, made possible by the discovery of new synthetic materials. W. Sterling Edwards and James Tapp invented the first artificial aortas, and as early as 1951, G. R. Denton had successfully replaced a mitral valve with an artificial valve in animal experiments.[79] By 1954, C. A. Hufnagel had placed an artificial heart valve in a human patient.[80]

Willem Kolff, inventor of the artificial kidney, began work in 1957 at the Cleveland Clinic on a totally artificial implantable heart.[81] His early model heart was based on a fuel pump from the White Motor Company. During the 1960s, surgeons Michael De Bakey and Denton Cooley began independent programs to develop an artificial heart. De Bakey had some success with a left-ventricle assist device (LVAD) which took over only part of the heart's function, keeping patients alive but still in a state of ill health. Somewhat in secret, and using some members of De Bakey's staff, Denton Cooley on April 14, 1969 became the first person to implant an artificial heart into a patient.[82] By 1971, Kolff had moved to the University of Utah where he was able to keep a calf alive for eleven days with an artificial heart. Part of the group assembled at the University of Utah was engineer-physician Robert Jarvik, who developed the ''Jarvik-7'' heart made from polyurethane and powered by a three hundred and seventy-five pound compressor. In December of 1982 surgeon William De Vries, another member of the Utah team, implanted a ''Jarvik-7'' heart into Barney Clark.[83] Clark lived for over four months, but the complication of strokes arising from blood clots formed in the heart indicated that more research and development needed to be done before an artificial heart was perfected. Recently, the artificial heart has been most successfully used as a bridge device to keep patients alive while awaiting transplants.

Researchers in biomedical engineering have made advances in several other areas. There has been a great deal of success in replacing knee and hip joints with materials such as metal or plastic.[84] Other work has been done on hearing aids that can be directly implanted into the ear as well as an artificial larynx. There has also been some success in developing surgically implantable capsules that would take the place of a particular gland.

All of these advances in biomedical engineering have raised questions concerning the distinction between the human and the machine. As each of these mechanical devices becomes a functioning part of a human, it becomes more and more difficult to characterize the assimilated object as a human or as a machine. In 1960 research space scientist Manfred Clynes coined the term *cyborg* (cybernetic organism) to refer to the new combination of human and machine.[85] The cyborg is not any ordinary combination of a human and a machine, such as a human using a tool; rather the cyborg involves a unique relationship between the human and the machine in that the machine "needs to function without the benefit of consciousness, in order to cooperate with the body's own autonomous homeostatic controls."[86]

Genetic Engineering

The assimilation of elements of the machine into the human body that resulted in the combination of the cyborg raised the more revolutionary possibility that biological organisms could be directly engineered. This idea arose out of research and discoveries in the field of genetics. In 1841 Robert Remark observed the fact that a cell was capable of producing another cell by dividing in half, and by 1855, Rudolf Virchow argued that all cells arose from pre-existing cells, but the detailed mechanisms of cell division were at that time unclear.[87] Further microscopic studies revealed some important details.[88] First, in 1861 Edouard-Gerard Babliani discovered that the cell nucleus contained chromosomes, a material that could easily be stained. Next, in 1875 Eduard Strasburger discovered that during cell division the nucleus of the cell also split into two parts. Finally, by 1884, a number of leading scientists began to recognize that chromosomes played an important role in cell division and might be the carriers of hereditary information. In 1900 the rediscovery of the work done by Gregor Mendel on the inheritance of dominant and recessive characters in peas, along with T. H. Morgan's later work with fruit flies helped to prove that chromosomes carry hereditary traits, such as eye color, in units called genes, which have specific chromosomal locations.

Although researchers in the early twentieth century were able to begin mapping the position of genes on the chromosomes, they still had very little knowledge of the chemical composition of the chromosomes or how they were able to regulate development. In 1869 Johann Fredrich Miescher began to study the chemical composition of the cell nucleus.[89] Using the large nuclei of white blood cells that he collected from the pus on surgical dressings, he was able to discover an acidic material composed of very large molecules. Since the material came from the cell nucleus, one of Miescher's students named it *nucleic acid*. Some of the basic chemistry of nucleic acid was worked out in the nineteenth century, and by 1920, two kinds of nucleic acid had been identified. The most common type contained a type of sugar called deoxyribose and was labeled *DNA* (deoxyribonucleic acid) while the other contained a type of sugar called ribose and was labeled *RNA* (ribonucleic acid). About the same time researchers, using a purple dye that would color only DNA, discovered that DNA was found only on the chromosomes.

Although DNA was located on the chromosomes, most biochemists thought that proteins were the key to understanding heredity. It was not until 1944 that Oswald Avery, Maclyn McCarthy, and Colin MacLeod were able to demonstrate that an ingredient that had been shown capable of transforming a nonvirulent pneumococcus bacterium into a virulent form acted more like DNA and not like a protein.[90] By 1952, Alfred Hershey and Martha Chase showed that when viruses, composed of DNA surrounded by a protein coat, infect a bacterium, only the DNA entered the bacterium.[91]

These studies convinced biochemists that the gene was composed of DNA and not proteins, but they still did not understand how the molecules of DNA were able to replicate themselves and pass on their traits to a new generation of cells. A major step toward an answer to the problem came in a book by the physicist Erwin Schrödinger, published in 1944, *What Is Life?*[92] Schrödinger saw the problem of life in terms of information and asked how a small amount of material in the nucleus of a fertilized egg, which was not yet identified as DNA, provided enough information for the organism's complete development. He suggested that the gene was a crystal composed of small units with the same form but whose atoms could be arranged in a variety of ways. Through these various arrangements, a relatively small number of atoms could provide an elaborate "code-script" similar to the Morse code. Schrödinger concluded, "What we wish to illustrate is simply that with the molecular picture of the gene it is no longer inconceivable that the miniature code should precisely correspond with a highly complicated and specific plan of development and should somehow contain the means to put it into operation."[93]

Schrödinger's book stimulated a new informationalist approach to genetics. A number of biologists, many of whom had moved to biology from physics, were especially influenced by Schrödinger's idea that the genes carried coded information. In order to discover how the genetic code functioned, however, the actual structure of DNA had to be determined. Although many people contributed to this enterprise, the problem was ultimately solved by James D. Watson and Francis Crick.[94] Both Watson, who was trained in biochemistry, and Crick, who was trained in physics, were strongly influenced by Schrödinger's book. Through a tremendous two-year undertaking involving a combination of X-ray analysis of a semicrystalline form of DNA, biochemical studies, and the informationalist approach, they were able to propose, in the April 1953 issue of *Nature*, that the DNA molecule was composed of a double helix in which two sugar-phosphate chains spiral in opposite directions held together, like the rungs of a ladder, by pairs of chemical compounds known as bases. The key to understanding DNA was that the four bases found in DNA, labeled *adenine* (A), *guanine* (G), *thymine* (T), and *cytosine* (C), fit together like a jigsaw puzzle so that adenine bonded only to thymine and guanine bonded only to cytosine. Thus the only pairs of bases allowed along the double helix were AT, TA, GC, and CG. This meant that the strands of the double helix formed a complementary image. Given the coded sequence of bases on any single strand of DNA, say CATG, the sequence on the other strand, GTAC could easily be constructed.

The Watson-Crick model of DNA gave researchers a new understanding of how genes replicate themselves as part of the hereditary process. In 1940, even before the double helix model, Linus Pauling and Max Delbrück suggested that genes might be able to duplicate themselves if they could produce a complementary molecule, similar to plaster of Paris, that could then produce copies of the original gene.[95] The double helix DNA could function in just such a way. If the two strands of a molecule of DNA were separated, or unzipped, each strand would be the complement of the other and each would provide a "template" for the formation of another strand of chemicals that would exactly reproduce the original DNA.

The double helix model of DNA opened up the possibility that scientists might be able to create, or at least manipulate, genetic material and thereby gain some control of the hereditary process.[96] Watson and Crick had used mechanical engineering terms such as *template* to describe the self-replication of DNA.[97] But before there could be any successful genetic engineering, researchers had to discover how DNA controlled the normal chemical activity that took place within a cell. For some time it had been recognized that the basic activity carried on within the cell was the production of proteins, long chainlike molecules composed of amino acids. Since the bases on a single strand of DNA could be in any order, researchers naturally began to speculate that there might be a connection between the sequence of bases in DNA and the sequence of amino acids that composed the proteins. The issue was complicated by the fact that DNA was found only in the nucleus of the cell while most protein production took place outside the nucleus in the cytoplasm. Therefore researchers began to focus their attention on RNA, which had a structure similar to DNA but was found both in the nucleus and in the cytoplasm. By the 1960s, researchers had discovered that DNA provided a template for the production of different kinds of RNA (messenger, transfer, and ribosomal) which were carried to the cytoplasm where they formed a template for the synthesis of amino acids into long protein chains. By 1966, they had also deciphered the genetic code; that is, they discovered how groups of three bases along the messenger RNA were associated with each of the twenty amino acids that made up the proteins.

The more researchers understood about DNA the more they began to use engineering models to describe its function. The original Watson-Crick model had been discussed in terms of "templates" and "codes."[98] Geneticist François Jacob, who won the 1965 Nobel Prize for Medicine, wrote that

> modern biology belongs to the new age of mechanism. The programme is a model borrowed from electronic computers. It equates the genetic material of an egg with the magnetic tape of a computer. It evokes a series of operations to be carried out, the rigidity of their sequence and their underlying purpose.[99]

The view of DNA as a biological computer program opened up the possibility that organisms could be created or changed by reprogramming or engineering their DNA.

This theoretical possibility became a reality in the late 1960s and early 1970s with the discovery of a new set of chemicals that became the basic tools of genetic engineering. It had been known for some time that certain enzymes had the ability to break the bonds of nucleic acids and cut them into smaller pieces. But these enzymes lacked precision and simply cut DNA into heterogeneous parts. By 1970, scientists such as Hamilton Smith began to discover some so-called "restriction enzymes" that cut DNA only at specific locations. Soon after a series of restriction enzymes were discovered that allowed scientists to make a wide range of specific cuts in the DNA chain. The fact that DNA could be cut precisely and certain sequences of bases isolated had revolutionary effects for the future of genetic engineering. Maxine Singer, one of the leading researchers in genetic engineering, has called these enzymes the basic "tools" that are needed for the manipulation of DNA.[100]

By 1973, all of the tools were in place to allow scientists to create a molecule of recombinant DNA (rDNA). That is, by cutting molecules of DNA with restriction enzymes, parts from two different DNA molecules could be "glued" together, using an enzyme called DNA ligase, to form a hitherto unknown DNA molecule. These molecules were actually created by several researchers including Paul Berg, Herbert Boyer, and Stanley Cohen. Their pioneering experiments meant that genetic engineering was a practical possibility. Within a few years Genentech became the first company founded to create drugs using the techniques of genetic engineering, and it was quickly followed by other companies—Biogen, Cetus, Genex, and Hybritech. Not only could these companies produce more easily such known drugs as insulin and interferon, they could also custom design previously unknown drugs and vaccines.

Although the complexity of DNA has limited most of the initial work on genetic engineering to relatively simple organisms, even this work had begun to raise questions about our long-held distinctions between what is organic and what is artificial. This dilemma was brought to a focus on June 16, 1980 when the United States Supreme Court, in a five-to-four vote, ruled that a new bacterium created by genetic engineering could be patented.[101] In 1972 Ananda Chakrabarty, a scientist working for the General Electric Company, had created, through cross-breeding, a bacterium that was capable of cleaning up an oil spill by digesting the oil. The original request by General Electric for a patent was turned down because of the view that "micro-organisms are 'products of nature,' and that as living things they are not patentable subject matter."[102] That decision was eventually appealed to the Supreme Court, and Chief Justice Warren Burger, writing for the majority, noted that the decision centered on the question of whether the new bacterium constituted a "manufacture" according to the law. The Court had previously ruled that "a new mineral discovered in the earth or a new plant found in the wild is not patentable" because there were "only some of the handiwork of nature."[103] But Burger argued that organisms produced by genetic engineering were different, "Here, by contrast, the patentee has produced a new bacterium with markedly different characteristics from any found in nature. . . . His discovery is not nature's handiwork but his own."[104] According to the Supreme Court decision, "the relevant distinction was not between living and inanimate things, but between products of nature, whether living or not, and human-made inventions.''[105]

With the development of genetic engineering, living organisms were no longer necessarily products of nature. Through the use of recombinant DNA it is possible to create entire new species by combining the DNA of two different species. Even more so than the cyborg, genetic engineering raises questions about the distinctions between the organic and the artificial. As with artificial intelligence, the organisms created by genetic engineering are the result of a new symbiosis between the organic and the technological and can best be understood as vital machines.

Conclusions: Convergence

Although computers and genetic engineering are usually seen as unrelated developments, both have evolved from the same bionic world view—neither can be fully understood in strictly mechanical or organic terms. By encompassing elements of both the organic and the mechanical in a dualistic system, each of these developments represents a path leading to the creation of a vital machine. And recognizing them as vital machines, we can begin to see an important interrelationship between the development of computers and genetic engineering. In his second edition of *Cybernetics* (1961), Weiner discussed how an electronic device might be designed that could be capable of causing another electronic device to reproduce the specific pattern of the original device.[106] His model for self-reproducing machines was taken from genetics. Weiner asks if ''this is philosophically very different from what is done when a gene acts as a template to form other molecules of the same gene from an indeterminate mixture of amino and nucleic acids?''[107]

If genes provided a model for the computer, the computer provided a model for genes. Nobel prize winning geneticist David Baltimore has written:

> Watson and Crick's model revealed that genes store information in a digital fashion; that the storage code has a four letter alphabet. . . . A possible implication, later proven, was that the letters are strung out along a virtually endless ''computer tape'' so that, although genes seem to be discrete entities, they are really just regions of the tape defined by encoded stop and start signals. . . . DNA could easily be seen as the embodiment of principles first realized in the late 1930s by Alan Turing and enshrined in the Turing machine, which responds to signals encoded on a tape.[108]

Recently work in artificial intelligence and genetic engineering has been moving toward a convergence that will be understandable only in terms of vital machines. Hans Moravec, an expert in robotics, has argued that we are entering a ''postbiological world,'' in which new forms of life, based on computers, will emerge.[109] As early as the 1950s, John von Neumann proposed a theory he labeled cellular automata, in which a set of cells, laid out in a checkerboard, could exist in different states, or colors, depending on some particular rule.[110] One rule might be that if a square cell was surrounded by at least three cells with the same color, the cell would take on that color. Using some very simple rules, a number of patterns could emerge. In some cases the patterns were simple, in others complex, and in still others stable patterns might "evolve," remain relatively stable for a period of time, and then "die" away. This theory led some computer scientists to see connections between cellular automata and life. In fact, mathematician John Conway has invented a set of rules for a cellular automaton called the Game of Life, which produces some patterns that destroy each other, and other patterns that yield a new generation of patterns.

By explicitly using ideas and concepts drawn from biology as rules for cellular automata, researchers are beginning to explore the possibility of *artificial life*. Biologist Richard Dawkins has used Darwinian principles to create a computer program in which small stick figures evolve based on a simplified genetic code, random mutations, and a few rules governing natural selection.[111] Christopher G. Langton at the University of Michigan used Edward O. Wilson's work on insect societies to create a colony of virtual ants, or "vants," which were programmed to cooperate in the task of building and following trails across the computer's screen.

Although such forms of artificial life are clearly distinct from actual organic life, recent developments are moving the two forms of life closer to one another. As early as the 1960s, computer programmers began to create, sometimes unintentionally, programs that could replicate themselves and become attached to other programs, causing those programs to do something they had not intended to do.[112] In the early days, when computers were few in number and programs had to be distributed on punched cards or tapes, these programs were of little consequence and usually created for amusement. But with the emergence of personal computers, mass-produced programs on floppy disks, computer networks, and electronic bulletin boards, these programs began to gain a great deal of attention.

In 1983 Fred Cohen, then a graduate student in computer science, developed the idea of a program that could copy itself into other programs; he showed that such a program could spread quickly throughout even secure computer networks.[113] Cohen labeled this program a virus since it functioned similarly to biological viruses by taking over control of the program it infects and using that program to make copies of itself to infect other programs. Any time an infected program was run, the virus would copy itself into the computer's memory and attach itself to the next program that was run on that computer. With the sharing of floppy disks among many computers and the linkage of computers into networks, a virus could quickly spread. Many computer viruses are benign such as the *MacMag* virus, which simply displayed a universal peace message on March 2, 1988 from the staff of *MacMag*, and then erased itself. But other viruses can be quite malicious. One called CRABS, invented at AT&T Bell Laboratories, produced small crablike images that began to eat away the material and data on the computer's screen.

Like biological viruses, computer viruses can be unpredictable and have a life of their own. In November 1988, Robert Tappan Morris, a graduate student at Cornell, designed a virus that was intended to be benign, but because of a programming error, it subsequently reproduced itself uncontrollably, shutting down Arpanet, the Department of Defense's national computer network. In her book *AIDS and Its Metaphors,* Susan Sontag notes a close connection between computer viruses and AIDS, "It is perhaps not surprising that the newest transforming element in the modern world, computers, should be borrowing metaphors drawn from our newest transforming illness. Nor is it surprising that the descriptions of the course of viral infection now often echo the language of the computer age."[114] Attempts at cures for computer viruses also reflect a close connection to biological disease. Some people have argued that one must avoid pirated copies of software, with unknown past histories, and limit the exchange of programs between computers—what has been called the practice of "safe hex."[115] Others have devised programs such as FLUSHOT, VACCINE, ANTIDOTE, ANTIGEN, and INTERFERON that search for viruses and eliminate them, or block the virus's ability to load itself into a program.[116]

Not all computer viruses or artificial life are destructive. Some computer scientists are using artificial life to create new computer programs that emerge through a kind of Darwinian evolution rather than by being designed by a person.[117] For example, Daniel Hillis at Thinking Machines has created a program to sort lists of numbers. His Connection Machine can test several different sorting programs simultaneously and combine the best programs into a new generation of programs. As the process is repeated, mutations are introduced in the form of random errors in the program. Most of these will cause a particular program to fail, but some will unexpectedly lead to a more successful program. To make sure there are a wide variety of programs from which to chose, viruses are added which weed out programs that might dominate the system, so that even more efficient programs might evolve.

As vital machines become more and more sophisticated, the distinction between artificial life and actual life may become moot. In the opinion of J. Doyne Farmer, a leading specialist in artificial life at the Los Alamos National Laboratory, "Although computer viruses are not fully alive, they embody many of the characteristics of life, and it is not hard to imagine computer viruses of the future that will be just as alive as biological viruses."[118] A development that could bring computer viruses closer to biological viruses is the possibility of creating computers using microchips that are composed of proteins rather than silicon.[119] Since many proteins exist in either of two states, they have the potential for use as binary language information processors. These "biochips" could be much smaller and faster than conventional silicon chips, and they could be produced using the techniques of genetic engineering. Using biochips, the action of a computer virus and the action of a biological virus might be difficult to distinguish.

If, as some scientists believe, a nanotechnology can be perfected, organisms and intelligent machines would become totally integrated producing the ultimate vital machine.[120] The term *nanotechnology* refers to machines on the scale of a nanometer (one billionth of a meter), the size of individual molecules. The possibility of such miniaturized machines was suggested as early as 1959 by physicist Richard Feynman, but only recently has the idea been given serious attention with the publication of K. Eric Drexler's paper on the subject in the *Proceedings of the National Academy of Sciences* and the formation of a Nanotechnology Study Group at MIT.[121]

According to Drexler, it should be possible to arrange organic molecules to form tiny motors, drive shafts, bearings, beams, and grasping arms. By using molecular rods with small molecular knobs attached, such as carbyne molecules that could slide between two positions, Drexler has suggested that molecular-sized computers could be designed. These elements could be combined to produce molecular-sized computer-controlled robots, called ''assemblers,'' that could bring together, or rearrange, individual atoms into some predetermined configurations. If such molecular robots were placed in the human body, they could move through the bloodstream, repairing cells by working on damaged segments of DNA. By programming one of the assemblers to assemble more robots, they could become self-replicating, and in great enough numbers they could be programmed to assemble anything, even large macroscopic objects, one atom at a time. Besides being of great potential benefit, such self-replicating molecular-sized robots, like computer viruses, could be devastatingly destructive. Carried by the air from place to place, they could act like biological viruses, with the exception that their damage would not be limited to organisms. They could ''infect'' anything that was composed of atoms.[122]

In a recent article novelist Thomas Pynchon has stated, ''If our world survives, the next great challenge to watch out for will come—you heard it here first—when the curves of research and development in artificial intelligence, molecular biology and robotics all converge.''[123] If nanotechnology becomes a reality, genetic engineering and artificial intelligence will have converged to produce vital machines.

8

Conclusion:
Ethics in the Age
of the Vital Machine

Technology is neither good nor bad, nor is it neutral.
Kranzberg's First Law

Traditional Ethics

The emergence of the vital machine raises a wide range of new ethical problems, and in order to solve many of these problems, ethics itself will have to undergo a significant transformation.[1] In the past ethics focused on human actions. Hans Jonas in his *Philosophical Essays* has given a brief characterization of the tacit assumptions of traditional ethical theories.[2] Traditional ethics was anthropocentric; only actions and relationships between human beings were of ethical significance. Also, the concept of human being that was at the center of ethics was assumed to be given and unchanging. Since human nature was a constant, the categories of good and evil could be easily determined. Most important, traditional ethics considered the nonhuman world, including organic nature and technology, as morally neutral, a given and a constant that could be little affected or changed by any human action. As Jonas has noted, "Action on non-human things did not constitute a sphere of authentic ethical significance."[3] Humans had little ethical responsibility for nature or technology; in statements of traditional ethics such as "Do unto others as you would have them do unto you," and "Love thy neighbor as thyself," the agent and the objects of moral concern are human beings.

In recent times, however, human action employing technology has been able to make a significant and long-term impact on the natural world. One result has been a demand for an ethics that can encompass our new responsibilities toward both nature and technology. According to Jonas, "Modern technology has introduced actions of such novel scale, objects, and consequences that the framework of former ethics can no longer contain them."[4] But recent attempts to extend ethical relationships to nature and technology have resulted in some deep divisions and even stalemates. Much of the divisiveness can be traced to the attempt of people in the twentieth century to continue to understand nature and technology in terms of the mechanical and organic world views.

From the perspective of the organic and mechanical world views, there is a fundamental conflict between the values associated with the organic and the mechanical. For example, according to the mechanical world view, the world, with the exception of human beings, is composed of a set of mechanical objects or things without any inherent spirit or vitality. These objects cannot have any wants or interests that need moral consideration.[5] The moral value of purely mechanical objects is determined by factors that are external to them—in effect, by their usefulness to human beings. According to this view, mechanical objects might have some instrumental value but they cannot have any intrinsic value. Technology is seen as the tools that are used to accomplish some humanly defined goal.[6] This humanly defined goal can be a subject of ethical discussion and judgments but the tools used to accomplish this goal have no intrinsic ethical status. The same knife can be used to carve a work of art or kill a human being. A rocket can deliver a nuclear weapon or allow us to explore the universe.

According to the mechanical world view, on the one hand, nature has the same limited status as technology. Since nature is simply a set of mechanical organisms without any inherent vitality, it also cannot have any wants or interests that need to be protected against exploitation. Like technology, nature can have instrumental value but not intrinsic value. That is, killing an animal or cutting down a tree might be morally wrong because these things belonged to someone else or because they would be more useful to more people if they were not killed or cut down. But they certainly would not be protected from harm because animals or trees had any inherent rights to exhibit elements of freedom or autonomy.

On the other hand, according to the organic world view, the world, including machines, is permeated with a spirit or vital force. Since all objects in the world have some kind of spirit or some level of consciousness, they all have some intrinsic needs that must be protected. Therefore the organic world view assumes that all objects in the world, not just human beings, have some intrinsic rights and that if an object has intrinsic rights it cannot be used as a mere instrument by some human being.[7] For example, St. Francis of Assisi argued that animals helped to glorify God in their own way, independent of their usefulness to human beings, and that they deserved ethical treatment based on their intrinsic rights to existence. In the sixteenth and seventeenth centuries some people argued that animals had rights because they, like human beings, were part of God's creation.[8] More recently, philosopher Peter Singer has advocated the idea of an animal liberation movement which argues that there are no circumstances, either for food or medical experiments, that justify the killings of animals.[9]

From the perspective of the organic world view, since technology encompasses a vital element, it has the same status as the objects of the natural world. Many modern critics argue that technology indeed has a life of its own.[10] Like Frankenstein's monster, an autonomous technology, although created by human beings, has taken on an independent freedom based on some internal force, and it develops in a direction that is no longer within human control.

The mechanical and organic world views have led to contradictory views concerning the role of technology and nature in modern ethical theories. From the perspective of the mechanical world view, both nature and technology have only instrumental value depending on their usefulness to human beings. In such a case our ethical concerns center on the issues of human responsibility for the use of technology and nature. But from the perspective of the organic world view, both nature and technology have intrinsic value. In this case our ethical concerns center on issues involving the freedom and autonomy of technology and nature. This contradiction between responsibility and control on the one hand and freedom and autonomy on the other is reflected in "new" ethical theories. If our relationship to technology is one of control and responsibility, it follows that nature has only instrumental value while if we give to nature the values of autonomy and freedom, it follows that we must lose control over our technology.[11] It is very difficult to conceive of an ethical theory that includes the values of control and responsibility as well as freedom and autonomy when we view the world as either mechanical or organic.

The breakdown of the mechanical and organic world views raises significant new questions about our ethical relationship to technology and nature. The ethical problems raised by the breakdown of a distinction between the organic and the mechanical can be seen in Karel Capek's famous play *R.U.R.* written in 1921. The title refers to Rossum's Universal Robots, a factory which produces robots (a term introduced by Capek derived from the Czech word *robota*, meaning forced labor). In the play, a famous physiologist named Rossum (from the Czech word *rozum*, meaning reason) has discovered a chemical synthesis that imitates living matter.[12] After limited success with using the substance to create an animal, Rossum attempts to create an artificial human, but ten years later he has only unsatisfactory results. He is then visited by his nephew, young Rossum, an engineer, who decides that artificial humans could be efficiently produced if they were greatly simplified, eliminating things such as the ability to play music or feel happiness. Young Rossum in fact creates robots, mechanical beings of great intellect but with no soul. A factory is established to produce young Rossum's robots to work as servants, factory workers, and soldiers. Problems begin when Dr. Gall, head of the physiological and experimental department of R.U.R., makes some changes in the robots that allow them to feel pain and experience human emotions. With these changes the robots begin to revolt against their human masters and start to kill them. Soon the robots supplant the humans who for some reason stop reproducing. Although the robots have been given souls by Dr. Gall, they do not know how to reproduce themselves since Rossum's original plans have been burned. At the end of the play the last surviving human witnesses the fact that two robots have fallen in love and will be the Adam and Eve of a new world.

The play depicts the emergence of the vital machine—mechanical robots with humanlike souls—as an attempt to overcome the inefficiency of organic life and the lack of spirituality of machines. The play is noteworthy today for raising several significant issues concerning the rights and ethical treatment of the emerging vital machine. If robots were considered as purely mechanical objects, there might be ethical problems concerning their instrumental value—their role in displacing humans from work or their role in war—but there would be no ethical problems concerning any intrinsic rights or their treatment by humans. If something went wrong with a robot, it could simply be sent to a stamping plant and destroyed. As long as the robots were viewed as purely mechanical, humans could exercise complete control and responsibility and the robots themselves would have no intrinsic rights.

But even without souls Rossum's robots had the physical appearance of humans, and within the play this fact begins a debate over the rights of the robots. Expressing an organic world view, a large group of characters in the play assemble to form the Humanity League, demanding that the robots be treated like human beings and be given their freedom. But from the mechanical world view, the idea of a robot being free and autonomous seems impossible. In responding to a demand that robots be given wages so that they can buy things they might want, an engineer for the factory responds, "That would be very nice . . . only there's nothing that does please the Robots. Good heavens, what are they to buy? You can feed them on pineapples or straw, whatever you like. It's all the same to them; they've no appetite at all. They've no interest in anything."[13] How can a mechanically controlled machine have any freedom?

Once Rossum's robots have been altered so that they have humanlike souls a new set of problems arises. Most significantly, the robots are no longer under the control of their human creators; their actions can no longer be predicted. Their new souls allow them the freedom to act in ways beyond the original designs of their human inventors. One of the robots states its new freedom: "I don't want a master. I want to be master."[14] Since the robots now have souls, the humans are forced to recognize a new relationship with the robots. As Dr. Gall states, "They've ceased to be machines."[15] Once the robots are no longer seen as purely mechanical inventions but as organic beings with some internal spirit or soul, humans have to accord them certain intrinsic rights. But these rights mean that the robots no longer have only instrumental value. At the end of the play one human, the architect Alquist, has been allowed to survive to help the robots discover the secrets of reproducing themselves. Since Rossum's original formula has been burned, the only solution is to dissect a robot to experimentally discover the formula. But Alquist says, "Am I to commit murder? See how my fingers shake! I cannot even hold the scalpel. No, no, I will not—."[16] If the robots are simply machines they could be destroyed but if they are "alive" they have intrinsic rights even if those rights lead to their own destruction and the destruction of humans.

The theme of *R.U.R.*, common in many other works of science fiction, can help us identify some of the ethical problems associated with the breakdown of the mechanical and organic world views. The two lines of traditional ethics based on these older world views both lead to unsatisfactory results when applied to the vital machine. If the robots in the play are treated as mechanical objects with only instrumental value, they cannot exhibit the autonomy and freedom of intelligence which is their distinguishing characteristic. A contradiction arises: the instrumental value of an intelligent machine cannot be simply instrumental. In the play some characters argue that the proper treatment of a machine like a robot is to liberate it.[17] But if the robots are treated as organic beings with intrinsic values, it becomes impossible for their human creators to take control of their creations. Once the robots are seen as having an autonomous soul or spirit, the contradiction of the intelligent slave arises. As the robots of *R.U.R.* become more intelligent they are no longer satisfied with being controlled by humans. They want to do things their own way.[18]

The ethics of the mechanical and organic world views lead to a contradiction when they are applied to the vital machine. In the sense that it is a machine it must be treated with an attitude of control and responsibility while in the sense that it is alive it must be treated with an attitude of autonomy and freedom. The emergence of the vital machine requires an ethical theory that can encompass the values of control and responsibility as well as the values of autonomy and freedom.

Ethical Problems in Artificial Intelligence

The recent development of actual vital machines in areas such as artificial intelligence and genetic engineering has forced people to face some of these ethical contradictions. For example, the development of artificial intelligence has raised some fundamental questions concerning the independence, autonomy, freedom, and rights of a humanly created object such as a machine. For many people the central issue of the ethical problems of artificial intelligence is the question of whether a machine, designed, planned, and constructed out of inorganic materials, can, in fact, exhibit behavior that we would associate with autonomy, freedom, and independence. That is, can machines be considered conscious beings?

During a symposium on the brain and machines, Satosi Watanabe of IBM addressed the issue on the basic level of the material difference between machines and life: "If a machine is made out of protein, then it may have consciousness, but a machine made out of vacuum tubes, diodes, and transistors cannot be expected to have consciousness."[19] But for others the fact that machines are composed of inorganic materials is not reason enough for rejecting the possibility of consciousness. The philosopher Hilary Putnam has argued that the problem of ascribing consciousness to machines or robots is analogous to the philosophical problem of determining the relationship between mind and body in human beings.[20] According to Putnam there is a philosophical problem in reducing a verbal statement that reports a mental state, such as "I am in pain," to a statement that reports a physical state, such as "a particular set of nerve fibers are being stimulated."[21] Although there might be a correlation between the stimulation of a particular set of nerve fibers and the mental state of experiencing pain, no one could argue that the nerve fibers themselves are experiencing pain. That is, mental states can be correlated to physical states but they cannot be identical to the physical state.[22] But if the mental states that we associate with human consciousness are "hopelessly different" from the physical state of the brain, there is no necessary reason why a robot composed of inorganic materials could not exhibit something analogous to a mental state. In fact Putnam shows that the printed statement of a Turing machine, "I am in logical state A," can be correlated with, but not identical to, the structural state of the machine. The machine could be composed of vacuum tubes, transistors, or mechanical switches and still be in logical state A.

Although Putnam's arguments do not prove that machines are conscious, they suggest the possibility that a robot could be considered "psychologically iso-morphic" to a human being.[23] That is, although a machine may be composed of mechanical and electronic parts, it still might obey the same rules of behavior as a human being. An intelligent machine may be materially and structurally distinct from humans, but if a "psychological parallelism" holds, we are faced with the problem that Putnam has labeled the "civil rights of robots,"[24] If intelligent machines exhibit behavior parallel to human consciousness, can we continue to treat them purely instrumentally or must we grant them some intrinsic rights?

Many arguments, which Putnam considers anti-civil libertarian, emphasize that a machine must behave in a deterministic and controlled manner. One of the oldest arguments against machine consciousness goes back to the nineteenth century. Alan Turing, in his article "Computing Machinery and Intelligence," called it "Lady Lovelace's Objection," after Lord Byron's daughter, who had written a memoir about Babbage's Analytical Engine in which she said that it "has no pretensions to *originate* anything. It can do *whatever we know how to order it* to perform."[25] According to this view, a machine only executes a series of predetermined operations and therefore cannot exhibit any free will or autonomy that we associate with consciousness. In recent times this objection has been put forward by Paul Ziff in an article "The Feelings of Robots."[26] According to Ziff, a robot can be programmed to behave as if it were tired, bored, or grieving but a robot cannot *feel* tired, bored, or grief stricken. A robot could be programmed to act tired after lifting a light weight while others could be programmed so that they never acted tired. As with a good actor, the outward behavior indicating grief reveals no proof of the inner feelings.

Putnam argues that these anti-civil libertarian arguments rely on the character-ization of robots as deterministic and preprogrammed to distinguish them from humans and to deny them any intrinsic rights.[27] But there are "pro-civil liber-tarian" arguments which raise problems concerning the characterization of robots as deterministic. Several researchers including Turing and Putnam question whether determinism is enough to distinguish robots from humans. Turing argued that all "original" activity, including that of humans, might be the result of some set of predetermined principles.[28] Others have contested that humans, through their DNA or through the set of physical forces we call Nature, are preprogrammed in ways similar to robots.[29] These same researchers and others have also claimed that Lady Lovelace's objection neglects the possibility that computers can be programmed to learn. Norbert Wiener used the example of a chess-playing machine to show that a computer would not be rigid and inflexible when playing against human opponents.[30] Using a high-order program, this machine could store in its memory previously played games and use the results of those games to revise the strategy that it was originally given so that it would create its own new strategies based on past experience. Although a programmer will have given the computer a set of rules, after the computer has played a number of games, that programmer will not know for certain what move the computer will make in a given situation. New developments in artificial life also raise questions about computers being predetermined.

More recently some scholars have raised newer versions of Lady Lovelace's objection. One of the most widely debated arguments for distinguishing artificial intelligence from human intelligence was proposed by philosopher Hubert Dreyfus in a paper ''Alchemy and Artificial Intelligence,'' and his later book *What Computers Can't Do: A Critique of Artificial Reason.* Dreyfus raises the question whether any form of artificial intelligence that must be reduced to a formal system of instructions, even those high-order programs that learn from past experience, can ever attain the same level of intelligence as human consciousness. As Lady Lovelace said, a computer can do ''whatever we know how to order it to perform,'' but can all human intelligence be formalized into a set of computer programs? Dreyfus notes that the major trend in Western thought, from Plato through Galileo, Leibniz, and down to Bertrand Russell, has been toward the idea that knowledge, as distinguished from belief, could be stated as explicit definitions or instructions that could be applied by anyone to the solution of a problem.[31] But Dreyfus argues for an alternative model of human knowledge, one that has been suggested by such twentieth-century philosophers as Edmund Husserl, Martin Heidegger, Maurice Merleau-Ponty, and Michael Polanyi. According to the theory of phenomenology, which is supported by many of these philosophers, much of our understanding about the world comes through intuition rather than through a rationality that reduces thought to some set of formal rules and operations. Dreyfus argues that certain elements of human thought, for example, riddles and other ill-defined games, language translation, recognition of distorted patterns or optical illusions, require types of thinking such as fringe consciousness and tolerance of ambiguity that are everyday human activities but are not governed by a formalized set of rules which can be imitated by a computer.[32]

Many scholars have responded to Dreyfus's critique by noting that some intuitive process may be able to be formalized.[33] Even Michael Polanyi, a philosopher often quoted by Dreyfus, admits that the intuitive process he calls ''tacit knowledge'' could be formalized.[34] Although the overly optimistic claims made by some researchers in the late 1950s have not yet come true, some progress has been made in the areas of translation, intuitive guesses, and context-oriented understanding, especially with the introduction of parallel processing. Dreyfus had argued that a computer could be taught to win at checkers but would never beat a world-class chess master. Several years ago Richard Greenblatt created a program called MacHack which beat Dreyfus at chess, and more recently a program called Deep Thought has beaten grand masters.[35] Many scholars have concluded that both sides in the Dreyfus debate may have some elements of the truth.[36] It may be unlikely that a computer will become an exact replica of human consciousness but this does not mean that computers will not exhibit some kind of intelligent behavior; however, the question of the civil rights of robots still must be addressed.

More recently another widely debated argument against attributing consciousness to computers was put forward by the philosopher John R. Searle in his article ''Minds, Brains, and Programs,'' in which he offers his ''Chinese Room Experiment.''[37] He proposes a thought experiment in which a person with no knowledge of Chinese is locked in a room and given different batches of Chinese writing along with a detailed set of rules, written in English, that allow the person to make correlations between elements in each batch of writing and to send out of the room certain Chinese symbols in response to certain symbols that are sent into the room. If the people outside the room call the symbols sent into the room "questions" and the symbols sent out of the room "answers," they would believe that the person inside the room "understands" Chinese. But Searle argues that the person inside the room does not "understand" Chinese. He is simply adept at manipulating squiggles according to the rules or "program" that has been given to him. Searle's "Chinese Room" is a rejection of Turing's test for artificial intelligence. According to Searle, even if a computer can behave exactly like a human mind by passing a Turing test for speaking Chinese, it cannot be conscious; it is simply a model or an imitation of consciousness.

Searle's argument caused a great debate among scientists, engineers, and philosophers over the consciousness of machines. His original article was published with twenty-eight responses. One of the most frequent responses Searle labeled *the systems reply,* and stated it as follows: "While it is true that the individual person who is locked in the room does not understand the story, the fact is that he is merely part of a whole system, and the system does understand the story."[38] That is, the understanding of Chinese cannot be attributed to the person in the room but to the room itself which includes the person along with the batches of writing and the detailed set of rules that the person uses to make correlations between "questions" and "answers." Searle responds by suggesting that the entire system could be adapted to the person in the room. If that person had a super memory, he could memorize all of the instructions and correlate the symbols in his head. According to Searle, the person still would not understand Chinese, but others question how we would distinguish this person from a "true" Chinese speaker. Do the individual neurons in the brain of a Chinese speaker "understand" Chinese any more than the person in Searle's room?[39]

The debate over the status of intelligent machines shows the contradictory arguments that arise when we try to argue that a vital machine is either a deterministic mechanical object with only instrumental value or an autonomous consciousness with intrinsic value. This dilemma has led some scholars to argue that we must simply "decide" how we are going to treat intelligent machines. Putnam says that the question of the civil rights of robots "calls for a decision and not for a discovery."[40] He says that if such a decision is made, "it seems preferable to me to extend our concept so that robots *are* conscious—for 'discrimination' based on 'softness' or 'hardness' of the body parts of a synthetic 'organism' seems as silly as discriminatory treatment of humans on the basis of skin color."[41] Others have argued that we must decide to limit our concept of consciousness to humans no matter how intelligent machines may become. Joseph Weizenbaum, the creator of the program ELIZA, has argued "that there are certain tasks which computers *ought* not be made to do, independent of whether computers *can* be made to do them."[42] He believes that artificial intelligence should never be substituted for a human function that involves "interpersonal respect, understanding, and love in the same category," and it should be avoided where it will lead to "irreversible and not entirely foreseeable side effects."[43]

But these agnostic approaches to artificial intelligence are not entirely satisfactory. Although they implicitly assume that neither the organic nor the mechanical world view is sufficient for dealing with intelligent machines, they nonetheless try to revert to one of those unitary world views by simply deciding to treat intelligent machines as either autonomous consciousnesses or mechanical objects. The recognition that the vital machine is the product of a new dualistic bionic world view can lead to a new way of understanding the ethical problems of conscious machines.

In her book *The Second Self,* sociologist and psychologist Sherry Turkle argues that many people cannot accept the idea of artificial intelligence because it involves a radically new and different concept of the self or the ego.[44] According to the traditional view, dating back to at least Descartes, thinking has been associated with an individual ''I'' or self. If there is thinking taking place, it must be located in a self or an ego. Searle rejects the notion that the room with its contents understands Chinese since the only possible location of thinking is the person in the room, who by definition does not understand Chinese. Like Searle, most of those who reject machine consciousness do so because they cannot locate a single agent or self among the mechanical parts of the machine.

The belief in a unitary localizable agent or self also leads to a fundamental distinction between values that are intrinsic, such as autonomy and freedom, and values that are instrumental, such as determinism and control. Values that are intrinsic are associated with the existence of a self or an ''I'' while values that are instrumental are associated with a being used by a self or an ''I'' for some purpose. If the self can be located only in certain things such as humans, then only those things can have intrinsic values while all other objects which do not have localizable selves, such as machines, can have only instrumental values.

But as Turkle discovered in her study, most scientists who support the idea of artificial intelligence reject the notion that thinking or consciousness must be associated with a single agent or ''I.''[45] Some of these scientists would deny that there is any self at all. For them thinking simply emerges from the sum total of the individual logical tasks accomplished by some material system, whether neurons of the brain or silicon chips in a computer. But many cannot totally eliminate the concept of the self. Rather than rejecting the self these scientists suggest a view of a ''decentered'' self or they ''recenter'' the self in such a way that it can inhabit both humans and machines. One computer science student, interviewed by Turkle, argued that the mind had two parts, computational and emotional, that coexisted. Another student argued that the self is divided between the hardware of the brain and the software that is the soul.[46] The self cannot be located in either the hardware or the software; rather it is decentered between both.

The concept of a ''decentered'' self can be traced back to Freud, whose theories of psychoanalysis arose out of a theory of the mind that combined a mechanical model of the brain with a vitalistic model of the psyche or soul. His formulation of the unconscious challenges the common sense idea that our concept of self must be associated with our conscious mind. Our desires and feelings are not controlled by the conscious ego but are hidden from us and repressed by the unconscious id. Thus Freud divided the self between the conscious and the unconscious. Although many of Freud's followers, including Anna Freud and Heinz Hartmann, placed renewed emphasis on the ego, more recently the French psychoanalyst Jacques Lacan has attacked ego psychology and argued that the ego was not a unified center but was constructed out of a confusion between the "I" and the "Other."[47] Some of Lacan's followers have seen the "decentered" self as a machine. In their book *Anti-Oedipus: Capitalism and Schizophrenia,* Gilles Deleuze and Félix Guattari have argued that the self is simply a collection of "desiring-machines," which never work together as a whole but interact with other "desiring-machines" in a completely fragmented way.[48]

Although the ethical debate over the civil rights of robots has not been solved, a study of it leads us to the conclusion that the new ethical theory of the vital machine will require a radically different framework than has existed in traditional theories of ethics. No longer can the focus of the theory of ethics be the autonomous individual. No longer can ethical judgments be based on a simple distinction between the intrinsic value of human beings and the instrumental value of technological creations. The focus of the ethics of the vital machine must be decentered. But this does not mean that there is no focus at all or that every object in the physical universe has an equal ethical status. The focus of responsibility, duty, and obligation must be in the symbiotic interactions that exist between human beings and their technological creations. In some cases, with the use of traditional tools, the interactions may be very simple and the "center" of ethics will be more on the side of the human, but in other cases, with the use of intelligent computers, the interactions may be quite complex and the "center" of ethics will be more or less equally divided between the human and the machine.

Recently passenger planes, such as the A320 built by the European consortium Airbus Industrie, have been designed to operate with "fly-by-wire" computers.[49] Instead of moving the flaps and elevators through a hydraulic system, the pilot uses a small joystick to send messages to a network of computers which control the plane. These computers have been programmed, with a Flight Envelope Protection System, to reject certain maneuvers that the airplane designers have decided are unsafe. In such a plane who is in control—the pilot or the computer—and who is responsible if something goes wrong? If a pilot tries to make a violent maneuver to avoid a collision and the computer will only allow a less violent turn or dive, who would be held accountable if a collision took place? Here the focus of responsibility would have to include both the pilot and the computer system—the human and the machine.

Ethical Problems in Genetic Engineering

A mirror image of the ethical problems concerning artificial intelligence has emerged in the debate over the dangers of future developments in genetic engineering.[50] The techniques of recombinant DNA have raised some fundamental questions concerning the ability of humans to control newly created or redesigned forms of life. Can the actions of living organisms, which have some level of independence and autonomy, be determined and controlled by humans who designed them? Can or should organisms which have some intrinsic value, by virtue of the fact that they are alive, be used in an instrumental fashion? That is, can life be *engineered?*

The ethical debate over genetic engineering began in the early 1970s when scientists began using animal viruses rather than plant viruses in some of their experiments. Paul Berg and his colleagues planned to place the chromosomes of an animal tumor virus (SV40) in a strain of bacterium called *Escherichia coli* (or *E. coli*), which was commonly used to drastically increase the original recombinant DNA molecule through its ability to quickly reproduce itself. But the *E. coli* bacterium used in the laboratory was closely related to the *E. coli* bacterium that was common in the human intestine and many researchers became worried that the genetically engineered DNA could spread to humans and cause a widespread outbreak of cancer or some other disease. Because of scientific concern Berg's experiment was postponed.

At 1973 at the annual Gordon Conference on Nucleic Acids, scientists began a serious discussion of the potential hazards of recombinant DNA. A letter published in *Science* by Maxine Singer and Dieter Soll, co-chairs of the conference, focused on the problem of controlling humanly designed organisms. They argued that the new experiments could result in "biological activity of unpredictable nature," and they called for the establishment of guidelines for experimentation with animal viruses.[51] Early the next year the National Academy of Sciences appointed a Committee on Recombinant DNA. Under the chairmanship of Paul Berg, the committee sent a letter to *Science* in 1974 asking for a voluntary moratorium on certain types of experiments, such as those involving animal viruses or new antibiotic resistance, and calling for an international meeting of scientists to discuss possible methods to control "the potential biohazards of recombinant DNA molecules."[52] This was one of the first times in history that scientists called for limitations on future research because of the potential dangers of that research.

In February 1975 a distinguished international group of scientists met at the California resort of Asilomar, under the sponsorship of the National Academy of Sciences, to debate the potential hazards of genetic engineering and to establish guidelines to control those hazards. A few scientists argued that no restrictions should be placed on scientific research, but the majority saw the need for some controls. The basic problem that had to be dealt with at Asilomar centered on the inherent contradiction of engineering genetic material. In an article in the *New Scientist* about the results of the conference, biologist Robert Sinsheimer summarized the problem of genetic engineering: "The essence of engineering is design and, thus, the essence of genetic engineering, . . . is the introduction of human design into the formulation of new genes and new genetic combinations."[53] But the human design of engineering had to face a new reality. "However, we must remember" Sinsheimer continued, "that we are creating here novel, self-propagating organisms. And with that reminder, another darker side appears on this brilliant scientific enterprise. . . . Abruptly we come to the potential hazard of research in this field."[54] That is, other forms of engineering such as nuclear reactors and chemical production may produce hazards, like radiation or chemical toxins, nevertheless, those hazards are proportional to the amount of radioactive or chemical materials that are released into the environment. The release of a single molecule of such material would not pose a widespread danger, but a single virus released into the environment, given the right circumstances, could reproduce to form an enormous number of viruses that could cause a worldwide epidemic.[55]

Because of the unique properties of biological organisms the scientists at Asilomar were faced with the problem of controlling something that was unpredictable in its behavior because of the intrinsic nature of the organism itself. For low-risk experiments on organisms that normally exchange genetic materials and on organisms that posed little threat to the environment, external controls could be established with physical containment using standard microbiological practices such as protective clothing, disinfectants, and safety cabinets.[56] But even the use of advanced physical containment techniques such as air locks, incineration of exhaust air, and negative pressure laboratories for chemical warfare research, could not eliminate the possibility that a single virus or bacterium could escape. Some other form of control was needed.

A few months before the conference, biologist Waclaw Szybalski wrote a letter to Paul Berg in which he suggested that organisms could be biologically contained using the intrinsic biology of such host organisms as *E. coli*. He said, "I strongly believe that clever genetic design is much superior to any costly and basically inefficient solutions based on 'brute force' mechanical containment systems."[57] Through the efforts of biologist Sydney Brenner, the Asilomar conference began to focus its attention on the use of "ecologically disabled organisms," or "self-destructing vectors," as a method of containing dangerous organisms. Recombinant DNA experiments require the use of a host organism, or vector, such as the *E. coli* bacterium. But the accidental spread of new DNA could be controlled if special bacteria or vectors could be designed that were unable to live outside of very special laboratory conditions. For example, it might be possible to create a strain of *E. coli* that could not multiply without a special nutrient found only within the laboratory, or unable to multiply in temperature ranges found in the natural environment.[58] Through the development of new strains of *E. coli*, nature itself, by the process of natural selection, would control the spread through the environment of potentially dangerous organisms. If the *E. coli*, used as hosts in laboratory experiments, contained a mutation that made its ability to reproduce occur less than the naturally occurring variety, then the doctrine of "survival of the fittest" would cause the natural variety of *E. coli* to overwhelm the laboratory version and make it extinct. The scientists at Asilomar realized that genetically engineered organisms could not be completely controlled using mechanically imposed physical containment. Rather, the control of new living material would have to take into consideration the vitalistic aspect of the organism itself. Although scientists would have to develop the new strains of *E. coli*, the ultimate control of the organisms could not be in human hands but would have to be contained by the biological force of natural selection.

Shortly after the Asilomar meeting the Recombinant DNA Molecule Program Advisory Committee (RAC) of the National Institutes of Health began to translate the recommendations of the scientists at Asilomar into a formal set of rules. By the end of 1975, RAC had created a proposed set of guidelines.[59] Experiments were classified according to their potential hazards and each level of hazard was matched to a category of containment procedures. Following the Asilomar recommendations, both physical and biological containments would be used. The most harmless experiments could be done in laboratories using normal micro-biological procedures (so-called P1 labs) and the standard laboratory strain of *E. coli* (so-called EK1), while the most dangerous experiments would have to done in laboratories using negative pressure air locks (P3), or in laboratories such as the Army's biological laboratory at Fort Detrick (P4) using strains of *E. coli* (EK2 or EK3) that would have an extremely small chance of living outside laboratory conditions (only one in one hundred million would be expected to survive). On June 23, 1976, the National Institutes of Health with the concurrence of the Secretary of Health, Education and Welfare issued formal guidelines based on the recommendations of the RAC.

Although scientists argued that biological containment was the ultimate method of control for genetic engineering, no "safe" bacteria (EK2 or EK3) had been certified for use at the time the guidelines were issued. The participants at Asilomar thought that it would take only a few weeks to develop a safe bacterium, but as one researcher said, "*E. coli* has enormous resiliency and a great capacity to do the illogical and unexpected."[60] Mutations in cells occur in very very small numbers, so it was difficult to isolate from the normal population the cells with the desired properties. Also, the mutations had to arise from the deletion of a part of the DNA in the bacterium, since other kinds of mutations that arise from the rearrangement of parts of the DNA could revert back to their normal state. The first safe *E. coli* bacterium was produced in 1976 by Roy Curtiss, III at the University of Alabama. The strain labeled *chi* 1776 after the Bicentennial, possessed sixteen different mutations such as the need for diaminopimelic acid, which would preclude it from existing in the human intestinal tract, and a mutation that gave it a fragile cell wall that would burst open in the presence of salt or detergent.[61] Soon after other safe hosts were developed and by the spring of 1977 they were certified by the National Institutes of Health for use in experiments requiring EK2 biological containment.

In recent times scientists have placed more and more confidence in biological containment. In June 1977 the National Institutes of Health and the National Institute of Allergy and Infectious Diseases sponsored a workshop at Falmouth, Massachusetts to study the basic biology of *E. coli* and to assess the risks that the bacterium might carry some known or unknown disease to humans in epidemic proportions. After studying the normal laboratory variety of *E. coli* (so-called K12) and the "safe" variety (*chi* 1776), the workshop concluded that although the K12 variety might survive outside the laboratory and might even survive in the human intestinal tract, it was impossible or extremely unlikely that K12 could cause an epidemic disease since it does not possess the factors that make normal *E. coli* virulent.[62] Based on the Falmouth workshop and other European studies, the National Institutes of Health guidelines were revised in December 1978 to allow some experiments that previously required "safe" *E. coli* (EK2) to be carried out with the normal laboratory strain K12.[63] Thus the inherent biology of K12 would control its own spread if it were released outside the laboratory.

The containment of new and potentially hazardous organisms did not totally resolve the ethical debate over genetic engineering. Beyond the immediate danger that recombinant DNA might cause an epidemic, several scientists, including Robert Sinsheimer, saw a potential long-term danger if genetic engineering began to interfere with and reorder the process of three billion years of evolution.[64] Through the use of recombinant DNA the barriers between species could be overcome and the long process of evolution would be irreversibly disturbed. These researchers believed that the only way to control the potential long-term hazards of "breaching the species barrier" would be to forbid such experiments.[65]

But other scientists believed that the forces at work in evolution itself would provide a kind of biological containment that would limit any long-term hazards of breaching the species barrier. Bernard Davis, a bacteriologist at Harvard Medical School, formulated the major response to Sinsheimer and others.[66] Davis argued that the critics who emphasized the dangers of recombinant DNA had focused on only one element of evolution. Since Darwin had postulated that hereditary variation was the essential basis of evolutionary change, an increase in that variation, brought about by genetic engineering, could change the future direction of evolution. Davis also noted that natural selection was an equally important element of evolution since it was "the main guiding force, determining which variants survive and spread."[67] According to this view, "the barriers that evolution has established between species are accordingly designed to avoid wasteful matings, that is, matings whose products would be monstrosities, unable to survive, rather than monsters, able to take over."[68] Therefore, genetically engineered organisms with DNA from different species would be unlikely to have the improved fitness needed for survival by means of natural selection. As with *E. coli,* scientists could count on a kind of biological containment to limit the spread of newly created organisms.

The debate over recombinant DNA is another example of the ethical problems associated with the creation of vital machines. The scientists working in the area of genetic engineering have come to recognize that the organisms that they have designed and are using have both instrumental value and an intrinsic autonomy and independence. From the perspectives of the organic and mechanical world views, the dualistic nature of genetic engineering can lead scientists to a dilemma concerning the ethics of recombinant DNA experiments. According to the mechanical world view, the organisms created by genetic engineering have only instrumental value. As such the organisms are seen simply as tools whose purpose and use can be externally controlled by their human creators. Based on this view, recombinant DNA organisms could be adequately controlled through physical containment, and no restrictions should be placed on the types of experiments that can be carried out.[69] But this view ignores the potential hazards arising from the intrinsic vitality of the organisms. According to the organic world view, the organisms created by genetic engineering have an intrinsic autonomy and indepen-dence that cannot be controlled by their human creators. Since the organisms cannot be controlled, very strict limitations should be placed on the types of experiments that scientists are allowed to carry out.[70] This view, however, ignores the instrumental value of recombinant DNA for producing new medicines and curing diseases.

Although the ethical debate over the hazards of recombinant DNA continues, most researchers have come to realize that genetic engineering cannot be treated from the perspective of either the mechanical or the organic world view. As with artificial intelligence, the debate over recombinant DNA requires a new ethical approach based on the concept of the vital machine. The focus of the debate can no longer be what scientists can do to organisms or what those organisms can do to us. Rather, the ethical debate over genetic engineering must focus on the symbiotic relationships that exist between ourselves and the genetic structure that controls organic life, including our own. The emergence of the idea of biological containment is an indication that scientists are beginning to recognize the dualistic nature of genetic engineering. The control of recombinant DNA is no longer solely in the hands of the scientists who can redesign the genes but neither is it solely in the genetic material itself. Actually the control resides in the interactions that exist between the internal design of the organism and the designs imposed on the organism by the scientists.

As with artificial intelligence, the symbiotic nature of genetic engineering also leads to a new "decentered" view of the individual. In writing about new discoveries that have arisen in the relationship between humans and other micro-organisms, Lewis Thomas has argued that it will

affect our view of ourselves as special entities, as selves in charge of our own being, in command of the earth. Another way to put it is that what we might be, in real life, is a huge collection of massive colonies of the most primitive motile life in air constructing around themselves, like a sort of carapace, all the embellishments and adornments of the human form.[71]

A Bionic Ethic

A study of the attempts to deal with the ethical problems raised by artificial intelligence and genetic engineering leads to the conclusion that a bionic world view will require a new ethical framework if it is to address problems raised by the emergence of the vital machine. Treating the relationship between the organic and the mechanical as a dualistic system, the ethical framework of the bionic world view must reject as its basis the concepts of an autonomous individual and a morally neutral technological world, but must instead be founded on the symbiotic interactions that exist between technology and organic life. Rather than centering on a machine or a human being, the new "decentered" focus of ethics must be on that dualistic mechanical-organic system that I have labeled the vital machine. The problem that we now face is not protecting the organic from mechanical inter-ference but recognizing that technology has become part of our environment, that "artificiality" is part of the "natural" condition of the world. The new question that faces us is this: How can we develop an ethical framework that will apply to the vital machine?

This need for a new framework is apparent when we consider the use of machines to maintain life.[72] The ultimate ethical problem here is, under what conditions should the machines cease to be used?—and this problem appears in a wide range of manifestations from the use of "high-tech" equipment in neonatal intensive care units for premature infants, to the use of kidney machines and artificial hearts for the chronically ill, to the use of respirators, feeding tubes, and pacemakers for patients in irreversible comas. Traditionally, ethics has focused on the individual patient. It was assumed that the determination of life or death could be made independent of the machines that were being used to support the patient. When the heart stopped the patient was declared dead and the machines were turned off.

But recent advances in medical technology have necessitated changes in the traditional view of an autonomous patient who is either alive or dead. With the use of mechanical respirators, electrical pacemakers, implanted feeding tubes, and other technology, the vital signs of a patient can be maintained for an indefinite period of time, raising the question of whether the machines together with the patient have "created" a new concept of life. Although there have been recent attempts to redefine death, such as the Harvard criteria that use the concept of whole brain death, these attempts still assume that the patient's life or death can be defined independent of the machines.[73] Expressing a traditional approach, ethicist Paul Ramsey has argued that such criteria are proposals "for rebutting the belief that machines or treatments are the patient."[74]

But such proposals will solve only temporarily the ethical dilemma brought about by new medical technology. The Harvard brain death criteria appear to solve the problem because we have no machines, at present, that can artificially maintain human brain function, or even maintain an electroencephalogram (EEG). If, or when, this technology is developed, how will we distinguish the machines or the treatment from the patient? We will have truly entered the bionic world of vital machines.

There will be no easy answers to such new developments, but confronting their possibility forces us to go beyond our traditional ethical framework. We will no longer be able to consider organic life independent of surrounding and supporting technology. Instead we must begin to develop a decentered ethical framework that reflects a bionic world view.

One step towards such a framework is the recent formulation of an environmental ethic. Although some elements of an environmental ethic can be traced to the nineteenth century, it was the work of Aldo Leopold, in his book *A Sand County Almanac* (1949), that pioneered the concept of a "land ethic."[75] Leopold argued that ethics had to extend from dealing with relations between individuals or between individuals and society, to dealing with relations between human beings and "soils, waters, plants, and animals, or collectively: the land."[76] His idea of a land ethic implied a shift away from anthropocentrism towards holism. He noted that "a land ethic changes the role of *Homo sapiens* from conqueror of the land-community to plain member and citizen of it."[77] Because all elements of the land-community were interdependent, Leopold argued: "A thing is right when it tends to preserve the integrity, stability, and beauty of the biotic community. It is wrong when it tends otherwise."[78]

While the land ethic of Leopold focuses on the organic, and in fact is usually interpreted as being in opposition to technology, it does provide a model for including both the organic and the mechanical into the expanding boundaries of a new ethic. In point of fact, Leopold often explained the interdependence of the biotic element of nature in terms of engine parts or wheels and cogs.[79] But the most radical element of his land ethic is its inclusion of nonliving things such as water, rivers, rocks, mountains, and land itself into the expanded boundaries of ethical concerns.

For those who accept Leopold's inclusion of nonliving things, the idea that technology could have some ethical standing is not entirely unthinkable. In his book *Should Trees Have Standing?—Toward Legal Rights for Natural Objects,* legal scholar Christopher D. Stone notes that inanimate objects such as trusts, corporations, banks, and even ships have long been seen by the courts as possessors of rights. In 1844 the courts refused to return to the owner a ship that had been used by pirates without the owner's knowledge, saying: "This is not a proceeding against the owner; it is a proceeding against the vessel for an offense committed by the vessel."[80] Stone himself noted that his argument that trees should be given legal rights could be used to advance the idea that such things as humanoids and computers might also have rights.

Other writers have suggested that landmark buildings should be treated in a way similar to endangered species.[81] That is, certain buildings, although products of technology rather than nature, deserve to be considered a part of our environment. Although the loss of the redwoods would lessen our world, so would the loss of the pyramids, Stonehenge, the Parthenon, the Great Wall of China, or Notre Dame Cathedral. Such an idea was recognized by the federal government when it passed the National Historic Preservation Act in 1966.

One final group of inanimate objects for which some rights are seen to exist are works of art. Objects of artistic creation, it is argued, have an intrinsic right to exist and to be treated with respect. Most people would consider it unethical to willfully neglect, alter, fake, or destroy an object that is classified as a work of art. One reason why we treat art objects in such a way is that they can be seen as the frozen thoughts of the artists who created them. By its containing the mind and the imagination of an artist, a work of art achieves an ethical status.

But, as in a painting or a sculpture, a machine or structure embodies the minds of the engineers who designed it. This is not a new notion, and it was well expressed in the nineteenth century in Nathaniel Hawthorne's short story "The Artist of the Beautiful."[82] The story concerns an inventor who creates a small mechanical butterfly indistinguishable from a real butterfly. When asked whether it is alive he answers, "Alive? Yes, . . . it may well be said to possess life, for it has absorbed my own being into itself; and in the secret of that butterfly, and in its beauty—which is not merely outward, but deep as its whole system—is repre-

154 / The Vital Machine

sented the intellect, the imagination, the sensibility, the soul of an Artist of the Beautiful!''[83] If we see technology as embodied thoughts of its designers, then we must give it some ethical standing.

Extending some kind of standing to technology as well as to the land, requires that we now think in terms of a bionic ethic rather than simply a land ethic. We must begin to envision a new sense of community, a "cybernetic ecology," to use a term from the poet Richard Brautigan. In his poem "All Watched Over by Machines of Loving Grace," he foresees a world where deer and pines exist alongside computers and electronics.[84] In our new community, both mechanical and organic elements have some claim to ethical treatment.

The idea of a cybernetic ecology does not imply that machines should be given equal standing with humans or with animals, plants, rivers, or mountains. Even within nature, there is a hierarchy of living things, with some species dominant over others. A fundamental element of any ecological system is the "food chain." Throughout the environment the continued survival of one species is dependent on its being able to eat (or in more general terms transfer energy from) another part of the environment. How and where various technologies fit along the "food chain" of a cybernetic ecology is still open to question; while it seems clear that technological elements should not be placed above human, can we say that all technologies should exist below the snail darter?

A cybernetic ecology can reconcile the contradictions between instrumental value and intrinsic value that arose from the old distinctions between the mechanical and the organic. Within an ecological system, all elements have some intrinsic value but because of the interdependence within the system every element also has some instrumental value for the rest of the system. Each part of the ecology has a certain degree of autonomy, but in the context of the system, each part plays some role in the control of the entire ecology.

A bionic ethic must take into consideration both the mechanical and the organic aspects of the cybernetic ecology in order to maintain the system's integrity, stability, diversity, and purposefulness. Neither the mechanical nor the organic can be allowed to bring about the extinction of the other. In the conclusion of his book, Leopold seemed to recognize that even a land ethic could not eliminate technology: "We are remodeling the Alhambra with a steam-shovel, and we are proud of our yardage. We shall hardly relinquish the shovel, which after all has many good points, but we are in need of gentler and more objective criteria for its successful use."[85]

In a cybernetic ecology both technology and organic life must be intelligently conserved. The entire system might be worse off without the peregrine falcon or the snail darter, but it also might be worse off without telecommunications and much of medical technology. On the other hand we might not want to conserve nuclear weapons or dioxin, but we might also be better off if the AIDS virus became extinct. In the end, we will build a new Jerusalem only if we can find a harmony between organic life and technology.

Notes

Chapter 1

1. Bill McKibben, *The End of Nature* (New York: Random House, 1989), p. 163.
2. See Paul Ramsey, *Ethics at the Edges of Life* (New Haven: Yale University Press, 1978).
3. For example, see Hans Moravec, *Mind Children: The Future of Robot and Human Intelligence* (Cambridge, Mass.: Harvard University Press, 1988).
4. Victor Turner, *From Ritual to Theatre: The Human Seriousness of Play* (New York: PAJ Publications, 1982), pp. 20–60.
5. Mary Douglas, *Purity and Danger: An Analysis of the Concepts of Pollution and Taboo* (London: Routledge & Kegan Paul, 1966), p. 96.
6. Ibid., p. 38.
7. Turner, *Ritual to Theatre*, p. 28.
8. For examples, see Frederick Turner, "Cultivating the American Garden: Toward a Secular View of Nature," *Harper's* 271 (August, 1985):45–52; Rachel Carson, *The Silent Spring* (Boston: Houghton Mifflin, 1962); and Jonathan Schell, *The Fate of the Earth* (New York: Avon, 1982).
9. For examples see, Carolyn Merchant, *The Death of Nature: Women, Ecology, and the Scientific Revolution* (San Francisco: Harper and Row, 1980); Theodore Roszak, *Where the Wasteland Ends: Politics and Transcendence in Postindustrial Society* (New York: Doubleday, 1972); Siegfried Giedion, *Mechanization Takes Command* (New York: W. W. Norton, 1948); and Lewis Mumford, *Technics and Civilization* (New York: Harcourt, Brace and World, 1934).
10. Leo Marx, "Reflections on the Neo-Romantic Critique of Science," in Gerald Holton and Robert S. Morison, eds., *Limits of Scientific Inquiry* (New York: W. W. Norton, 1979), pp. 63–66.
11. Franklin L. Baumer, *Modern European Thought: Continuity and Change in Ideas, 1600–1950* (New York: Macmillan, 1977), pp. 268–301.
12. Alfred North Whitehead, *Science and the Modern World* (New York: The Free Press, 1925), p. 94.
13. Baumer, *Modern European Thought*, pp. 281–82.
14. Henry Petroski, "H. D. Thoreau, Engineer," *American Heritage of Invention and Technology* 5 (Fall, 1989):8–16.
15. Quoted in Marx, "Neo-Romantic Critique," p. 63.
16. See M. H. Abrams, *The Mirror and the Lamp: Romantic Theory and the Critical Tradition* (Oxford: Oxford University Press, 1953).
17. Ibid., pp. 159–67.

18. Ibid., pp. 167–77.
19. Quoted in ibid., p. 168.
20. Ibid., p. 169.
21. Quoted in ibid., p. 174.
22. See Leo Marx, *The Machine in the Garden: Technology and the Pastoral Ideal in America* (New York: Oxford University Press, 1964), pp. 169–90; and Marx, "Neo-Romantic Critique," pp. 64–66.
23. Thomas Carlyle, "Sign of the Times," in *The Works of Thomas Carlyle*, vol. 14 (New York: Peter Fenton Collier, 1897), p. 465.
24. Ibid., pp. 465–66.
25. Ibid., p. 466.
26. Ibid., pp. 466–67; and Marx, *Machine in the Garden*, p. 171.
27. Carlyle, "Sign of the Times," p. 468.
28. Ibid., p. 474.
29. Ibid., p. 475.
30. Ibid., pp. 471–72.
31. See Stephen C. Pepper, *World Hypotheses* (Berkeley: University of California Press, 1942), pp. 84–114; and George Lakoff and Mark Johnson, *Metaphors We Live By* (Chicago: University of Chicago Press, 1980).
32. Marx, *Machine in the Garden*, p. 4.
33. Howard Gardner, *Art, Mind, and Brain: A Cognitive Approach to Creativity* (New York: Basic Books, 1982), p. 5.
34. Ernst Cassirer, *Language and Myth*, trans. Susanne Langer (New York: Dover, 1946), p. 7.
35. Ernst Cassirer, *An Essay on Man: An Introduction to a Philosophy of Human Culture* (New Haven: Yale University Press, 1944), p. 26.
36. Ibid., p. 25.
37. Susanne K. Langer, *Philosophy in a New Key: A Study in the Symbolism of Reason, Rite, and Art*, 3rd ed. (Cambridge: Harvard University Press, 1982), p. 26.
38. Ibid., pp. 26–79.
39. Ibid., pp. 60–61.
40. Clifford Geertz, *The Interpretation of Cultures* (New York: Basic Books, 1973), pp. 126–41.
41. Ibid., p. 127.
42. Ibid.
43. Jerome Bruner, *Actual Minds, Possible Worlds* (Cambridge: Harvard University Press, 1986), pp. 93–105.
44. Nelson Goodman, *Ways of Worldmaking* (Indianpolis: Hackett, 1978), p. 6.
45. See Hilde Hein, "The Endurance of the Mechanism–Vitalism Controversy," *Journal of the History of Biology* 5 (1972):159–88.

Chapter 2

1. For example, see Otto Mayr, *Authority, Liberty and Automatic Machinery in Early Modern Europe* (Baltimore: Johns Hopkins University Press, 1986), Ch. 3; Hilde Hein, "The Endurance of the Mechanism–Vitalism Controversy," *Journal of the History of Biology* 5 (1972):159–88; Robert E. Schofield, *Mechanism and Materialism: British Natural Philosophy in An Age of Reason* (Princeton: Princeton University Press, 1969); Marie Boas, "The Establishment of the Mechanical Philosophy," *Osiris*

10 (1952):412–541; and E. J. Dijksterhius, *The Mechanization of the World Picture*, trans. C. Dikshoorn (Oxford: Oxford University Press, 1961).

2. See Marshall Clagett, *Greek Science in Antiquity* (New York: Collier, 1955); David Furley, *The Greek Cosmologists: Volume 1: The Formation of the Atomic Theory and Its Earliest Critics* (Cambridge: Cambridge University Press, 1987); and G. E. R. Lloyd, *Early Greek Science, Thales to Aristotle* (New York: W. W. Norton, 1970).

3. Stephen Toulmin and June Goodfield, *The Architecture of Matter* (Chicago: University of Chicago Press, 1962), p. 55.

4. See Plato's *The Republic,* Book 7.

5. See Thomas Kuhn, *The Copernican Revolution* (New York: Vintage, 1957), Ch. 1; Mayr, *Authority, Liberty and Automatic Machinery,* pp. 1–4; and Clagett, *Greek Science.*

6. Mayr, *Authority, Liberty and Automatic Machinery,* p. 58.

7. Kuhn, *Copernican Revolution,* pp. 79–81; and Edward Grant, "Celestial Orbs in the Latin Middle Ages," *Isis* 78 (1987):153–73. Grant shows that the spheres could be fluid or rigid.

8. For a discussion of medieval technology, see J. Gimpel, *The Medieval Machine: The Industrial Revolution of the Middle Ages* (New York: Holt, Rinehart & Winston, 1976); Carlo M. Cipolla, *Before the Industrial Revolution: European Society and Economy, 1000–1700* (New York: W. W. Norton, 1976); Lynn White, Jr., *Medieval Technology and Social Change* (Oxford: Oxford University Press, 1962); and Derek J. de Solla Price, "Automata and the Origins of Mechanism and Mechanistic Philosophy," *Technology and Culture* 5 (1964):10.

9. For a history of clocks, see David S. Landes, *Revolution in Time: Clocks and the Making of the Modern World* (Cambridge, Mass.: Belknap Press, Harvard University Press, 1983); Carlo M. Cipolla, *Clocks and Culture, 1300–1700* (New York: W. W. Norton, 1978); and Mayr, *Authority, Liberty and Automatic Machinery.*

10. Mayr, *Authority, Liberty and Automatic Machinery,* pp. 3–6.

11. Landes, *Revolution in Time,* p. 53.

12. Mayr, *Authority, Liberty and Automatic Machinery,* Ch. 2.

13. Cipolla, *Clocks and Culture,* pp. 103–7.

14. Ibid., p. 45.

15. Quoted in Mayr, *Authority, Liberty and Automatic Machinery,* p. 26.

16. Robert H. Kargon, *Atomism in England from Hariot to Newton* (Oxford: Clarendon Press, 1966), pp. 1–17; and Christoph Meinel, "Early Seventeenth-Century Atomism: Theory, Epistemology, and Insufficiency of Experiment," *Isis* 79 (1988):68–103.

17. See Schofield, *Mechanism and Materialism.*

18. Boas, "The Establishment of the Mechanical Philosophy," p. 438.

19. Pietro Redondi, *Galileo: Heretic,* trans. Raymond Rosenthal (Princeton: Princeton University Press, 1987). Redondi argues that the trial of Galileo was over his atomistic beliefs which were a threat to the doctrine of transsubstantiation. Also see, Stillman Drake, *Galileo at Work: His Scientific Biography* (Chicago: University of Chicago Press, 1978); A. Mark Smith, "Galileo's Theory of Indivisibles: Revolution or Compromise?" *Journal of the History of Ideas* 37 (1976):571–88; William R. Shea, "Galileo's Atomic Hypothesis," *Ambix* 17 (1970):13–27; and E. A. Burtt, *The Metaphysical Foundations of Modern Science* (Garden City, N.Y.: Doubleday Anchor, 1954), pp. 72–104.

20. Quoted in Burtt, *Metaphysical Foundations,* p. 75.

21. See Boas, "Mechanical Philosophy," p. 435; and Burtt, *Metaphysical Foundations,* p. 87.

22. Galileo Galilei, *Dialogues Concerning Two New Sciences,* trans. Henry Crew and Alfonso de Salvio (New York: Dover Publications, 1954), p. 18.

23. Ibid., p. 19.

24. Boas, "Mechanical Philosophy," pp. 435–36.

25. Galileo, *Two New Sciences,* p. 42.

26. Quoted in Burtt, *Metaphysical Foundations,* p. 88.

27. René Descartes, "Discourse on the Method of Rightly Conducting the Reason and Seeking for Truth in the Sciences," in *The Philosophical Works of Descartes,* vol. 1, trans. Elizabeth Haldane and G. R. T. Ross (New York: Dover Publications, 1955), p. 85.

28. Ibid., p. 92.

29. Ibid., p. 101.

30. Ibid.

31. Richard S. Westfall, *The Construction of Modern Science, Mechanisms and Mechanics* (Cambridge: Cambridge University Press, 1977), p. 31.

32. Descartes, "The Principles of Philosophy," in *Philosophical Works,* vol. 1, pp. 255–56.

33. Kargon, *Atomism in England,* p. 64.

34. Westfall, *Construction of Modern Science,* p. 33.

35. Ibid.

36. Ibid., pp. 36–37.

37. Quoted in Burtt, *Metaphysical Foundations,* p. 119.

38. See Barry Brundell, *Pierre Gassendi: From Aristotelianism to a New Natural Philosophy* (Dordrecht: D. Reidel, 1987); Kargon, *Atomism in England,* pp. 65–68; Westfall, *Construction of Modern Science,* pp. 39–42; and Boas, "Mechanical Philosophy," pp. 429–31.

39. Kargon, *Atomism in England,* p. 67.

40. See ibid.; and Boas, "Mechanical Philosophy," p. 430.

41. Kargon, *Atomism in England,* p. 68.

42. See ibid.; Francis Bickley, *The Cavendish Family* (Boston: Houghton Mifflin, 1914); Helen Hervey, "Hobbes and Descartes in Light of Some Unpublished Letters of the Correspondence between Sir Charles Cavendish and Dr. John Pell," *Osiris* 10(1952):65–90; and Jean Jacquot, "Sir Charles Cavendish and His Learned Friends," *Annals of Science* 8(1952):13–27, 175–91.

43. See G. R. de Beer, "Some Letters of Thomas Hobbes," *Notes and Records of the Royal Society* 7(1950):195–206; Charles T. Harrison, "Bacon, Hobbes, Boyle, and the Ancient Atomists," *Harvard Studies and Notes in Philology and Literature,* vol. 15 (Cambridge: Harvard University Press, 1933); Thomas Hobbes, *The Metaphysical System of Hobbes* (LaSalle, Ill.: Open Court Publishing Co., 1963); Jean Jacquot, "Notes on an Unpublished Work of Thomas Hobbes," *Notes and Records of the Royal Society* 9(1952):188–95; and Steven Shapin and Simon Schaffer, *Leviathan and the Air-Pump: Hobbes, Boyle, and the Experimental Life* (Princeton: Princeton University Press, 1985).

44. Kargon, *Atomism in England,* pp. 56–57.

45. Quoted in ibid., p. 56.

46. Ibid., pp. 56–58.

47. Thomas Hobbes, *The English Works of Thomas Hobbes,* vol. 1 (London: John Bohn, 1839), p. 474.

48. Ibid., p. 426.

49. Ibid., p. 476.

50. Ibid., vol. 7, p. 32.

51. Ibid., vol. 1, p. 477.
52. Ibid., vol. 7, p. 35.
53. Ibid., vol. 1, p. 478.
54. See ibid., vol. 1, p. 476; and vol. 7, pp. 34–35, 134–35.
55. Kargon, *Atomism in England,* pp. 60-66.
56. Quoted in ibid., p. 61.
57. Quoted in Burtt, *Metaphysical Foundations,* pp. 126–27.
58. Quoted in Kargon, *Atomism in England,* p. 61.
59. See ibid., pp. 93–105; Burtt, *Metaphysical Foundations,* pp. 162–207; and Thomas Kuhn, "Robert Boyle and Structural Chemistry in the Seventeenth Century," *Isis* 43(1952):12–36.
60. See J. E. McGuire, "Boyle's Conception of Nature," *Journal of the History of Ideas* 33 (1972):523–42; and Kargon, *Atomism in England,* pp. 97–100.
61. Steven Shapin, "The House of Experiment in Seventeenth-Century England," *Isis* 79 (1988):373–404; and Shapin and Schaffer, *Leviathan and the Air-Pump.*
62. Quoted in Kuhn, "Robert Boyle," p. 17.
63. See Kargon, *Atomism in England,* pp. 102–5; and Westfall, *Construction of Modern Science,* pp. 75–81.
64. Quoted in Kargon, *Atomism in England,* p. 103.
65. Westfall, *Construction in Modern Science,* p. 77.
66. Burtt, *Metaphysical Foundations,* pp. 194–202.
67. Quoted in ibid., p. 166.
68. Quoted in ibid., p. 202.
69. Westfall, *Construction of Modern Science,* p. 140–41.
70. Ibid., pp. 127–28.
71. See Richard Westfall, *Never at Rest: A Biography of Isaac Newton* (Cambridge: Cambridge University Press, 1980); Richard Westfall, *Force in Newton's Physics* (London: Macdonald, 1971); I. B. Cohen, *The Newtonian Revolution: With Illustration of the Transformation of Scientific Ideas* (Cambridge: Cambridge University Press, 1980); I. B. Cohen, *Franklin and Newton* (Cambridge: Harvard University Press, 1966); and Alexandre Koyré, "The Significance of the Newtonian Synthesis," in *Newtonian Studies* (Chicago: University of Chicago Press, 1965), pp. 3–24.
72. Schofield, *Mechanism and Materialism,* p. 7.
73. Isaac Newton, *Mathematical Principles of Natural Philosophy,* vol. 1, trans. Florian Cajori (Berkeley: University of California Press, 1971), p. 5.
74. Isaac Newton, *Opticks* (New York: Dover Publications, 1952), pp. 375–76.
75. Ibid., p. 376.
76. Ibid., p. 377.
77. Schofield, *Mechanism and Materialism,* p. 11.
78. See B. J. T. Dobbs, *The Foundation of Newton's Alchemy* (Cambridge: Cambridge University Press, 1975); Margaret Jacob, *The Newtonians and the English Revolution* (Ithaca: Cornell University Press, 1976); Ernan McMullin, *Matter and Activity in Newton* (Notre Dame, Ill.: University of Notre Dame Press, 1977); J. E. McGuire, "Force, Active Principles, and Newton's Invisible Realm," *Ambix* 15 (1968):154–208; and J. E. McGuire, "Neoplatonism and Active Principles: Newton and the *Corpus Hermeticum,*" in Robert S. Westman and J. E. McGuire, eds., *Hermeticism and the Scientific Revolution* (Los Angeles: University of California Press, 1977).
79. See Alexandre Koyré, *From the Closed World to the Infinite Universe* (Baltimore: Johns Hopkins University Press, 1957), p. 237.
80. Newton, *Opticks,* p. 403.
81. Ibid., p. 402.

82. Ibid., p. 403.
83. Quoted in Koyré, *Closed World*, p. 236.
84. See ibid., pp. 235–72; A. Koyré and I. B. Cohen, "Newton and the Leibniz–Clarke Correspondence," *Archives Internationales d'Histoire des Sciences* 15 (1962):63–126; A. R. Hall and Marie Boas Hall, "Clarke and Newton," *Isis* 52 (1961):583–85; and Steven Shapin, "Of Gods and Kings: Natural Philosophy and Politics in the Leibniz–Clarke Disputes," *Isis* 72 (1981):187–215.
85. K. Maurice and Otto Mayr, *The Clockwork Universe: German Clocks and Automata, 1550–1650* (New York: Neale Watson, 1980).
86. See Schofield, *Mechanism and Materialism*, p. 134; and Burtt, *Metaphysical Foundations*, p. 298.
87. Koyré, *Closed World*, p. 276.
88. Schofield, *Mechanism and Materialism*, pp. 277–97.
89. See Arnold Thackray, *Atoms and Powers: An Essay on Newtonian Matter-Theory and the Development of Chemistry* (Cambridge: Harvard University Press, 1970); Alan Rocke, *Chemical Atomism in the Nineteenth Century* (Columbus: Ohio State University Press, 1984); Mary Jo Nye, "The Nineteenth-Century Atomic Debates," *Studies in History and Philosophy of Science* 7 (1976):245–68; and Aaron J. Ihde, *The Development of Modern Chemistry* (New York: Harper and Row, 1964), pp. 89–124.
90. See Arnold Thackray, *John Dalton: Critical Assessments of His Life and Science* (Cambridge: Harvard University Press, 1972); and D. S. L. Cardwell, *John Dalton and the Progress of Science* (Manchester: Manchester University Press, 1968).
91. See D. S. L. Cardwell, *From Watt to Clausius: The Rise of Thermodynamics in the Early Industrial Age* (Ithaca: Cornell University Press, 1971); Yehuda Elkana, *The Discovery of the Conservation of Energy* (Cambridge: Harvard University Press, 1974); and Thomas Kuhn, "Energy Conservation as an Example of Simultaneous Discovery," in Marshall Clagett, ed., *Critical Problems in the History of Science* (Madison: University of Wisconsin Press, 1969), pp. 321–56.
92. Schofield, *Mechanism and Materialism*, Part 2.
93. Franklin's positive and negative are the opposite of modern terminology. For information on Franklin, see Cohen, *Franklin and Newton*.
94. See Lancelot Law Whyte, ed., *Roger Joseph Boscovich, S.J., F.R.S., 1711–1787* (London: George Allen & Unwin, 1961).
95. For example, see Carolyn Merchant, *The Death of Nature: Women, Ecology, and the Scientific Revolution* (San Francisco: Harper and Row, 1980); C. D. Broad, *Mind and Its Place in Nature* (London: Routledge, Kegan Paul, 1923); and Schofield, *Mechanism and Materialism*, p. 95.

Chapter 3

1. Stephen Toulmin and June Goodfield, *The Architecture of Matter* (Chicago: University of Chicago Press, 1977), p. 309.
2. For a history of automata, see Derek J. de Solla Price, "Automata and the Origins of Mechanism and Mechanistic Philosophy," *Technology and Culture* 5(1964):11; and Alfred Chapius and Edmond Droz, *Automata* (New York: Central Book Co., 1958).
3. Price, "Automata," p. 11.
4. Marie Boas Hall, ed., *The Pneumatics of Hero of Alexandria* (New York: Science History Publications, 1971).
5. Ibid., pp. 62, 95, 111; and Chapius and Droz, *Automata*, pp. 34–35.

Notes / 161

6. Chapius and Droz, *Automata,* pp. 36–40.
7. Price, "Automata," p. 15.
8. "Jack" may be derived from the word *Jaccomarchiadus,* which meant man in a suit of armor. In fact the first clocks were made by armorers. See Jack Burnham, *Beyond Modern Sculpture* (New York: George Braziller, 1968), p. 195.
9. Chapius and Droz, *Automata,* p. 51.
10. See Burnham, *Modern Sculpture,* p. 193; and Price, "Automata," p. 19.
11. He believed that man had a rational soul and therefore was distinct from animals. See St. Thomas Aquinas, *Summa Theologica* (Qu. 13. Art. 2, Reply obj. 3, Pt. 2); and Price, "Automata," p. 19.
12. Frances Yates, *Theatre of the World* (Chicago: University of Chicago Press, 1969), pp. 31–32.
13. See Chapius and Droz, *Automata,* p. 40.
14. See Walter Pagel, *William Harvey's Biological Ideas* (New York: Hafner, 1967); and Robert G. Frank, Jr., *Harvey and the Oxford Physiologists: A Study of Scientific Ideas and Social Interactions* (Berkeley: University of California Press, 1981).
15. Pagel, *William Harvey,* p. 52.
16. Quoted in ibid., p. 211. There is some indication that this portion of the lecture dates from 1618. See George Basalla, "William Harvey and the Heart as a Pump," *Bulletin of the History of Medicine* 36 (1962):468n.
17. Basalla, "William Harvey," pp. 467–70.
18. Quoted in C. Webster, "William Harvey's Conception of the Heart as a Pump," *Bulletin of the History of Medicine* 39 (1965):510.
19. Ibid., p. 515.
20. Pagel, *William Harvey,* pp. 80–81.
21. See Richard B. Carter, *Descartes' Medical Philosophy: The Organic Solution to the Mind-Body Problem* (Baltimore: Johns Hopkins University Press, 1983); Ann Wilbur Mackenzie, "A Word about Descartes' Mechanistic Conception of Life," *Journal of the History of Biology* 8 (1975):1–13; and T. S. Hall, "Descartes' Physiological Method: Position, Principles, Examples," *Journal of the History of Biology* 3 (1970):53–79.
22. René Descartes, "Discourse on the Method," in *The Philosophical Works of Descartes,* vol. 1, trans. Elizabeth Haldane and G. R. T. Ross (New York: Dover Publications, 1955), pp. 115–16.
23. René Descartes, "The Passions of the Soul," in ibid., p. 333.
24. Descartes, "Discourse," pp. 112–15.
25. Leonora Cohen Rosenfield, *From Beast-Machine to Man-Machine: Animal Soul in French Letters from Descartes to La Mettrie,* new ed. (New York: Octagon, 1968), p. 6.
26. Quoted in *Selected Readings in the History of Physiology,* 2d ed., comp. John F. Fulton, completed by Leonard G. Wilson (Springfield, Ill.: Charles C Thomas, 1966), pp. 261–62.
27. See Julian Jaynes, "The Problem of Animal Motion in the Seventeenth Century," *Journal of the History of Ideas* 31 (1970):291–34; Wallace Shugg, "The Cartesian Beast Machine in English Literature (1663–1750)," *Journal of the History of Ideas* 29 (1968):279–92; Leonora D. Cohen, "Descartes and Henry More on the Beast Machine: A Translation of their Correspondence Pertaining to Animal Automation," *Annals of Science* 1 (1936):48–61; and Rosenfield, *From Beast-Machine,* pp. 3–26.
28. Rosenfield, *From Beast Machine,* pp. 37–40.
29. Ibid., pp. 73–107.

30. See Luigi Belloni, "Marcello Malpighi and the Founding of Anatomical Micro-scopy," in M. L. Righini Bonelli and William R. Shea, eds., *Reason, Experiment, and Mysticism in the Scientific Revolution* (New York: Science History Publications, 1975), pp. 95–98; and Francois Dushesneau, "Malpighi, Descartes and the Epis-tomological Problems of Iatromechanism," in ibid., pp. 111–16.

31. Quoted in Belloni, "Malpighi," pp. 96–97.

32. *Dictionary of Scientific Biography*, s.v. "Borelli, Giovanni."

33. See Howard B. Adelmann, *Marcello Malpighi and the Evolution of Embryology*, vol. 1 (Ithaca: Cornell University Press, 1966), p. 150.

34. See Richard S. Westfall, *The Construction of Modern Science: Mechanisms and Mechanics* (Cambridge: Cambridge University Press, 1975), pp. 94–96.

35. Quoted in Fulton and Wilson, *Selected Readings*, pp. 220–21.

36. For background on Malpighi, see Adelmann, *Malpighi*.

37. Belloni, "Malpighi," p. 95.

38. Ibid., p. 97.

39. Ibid., pp. 102–3.

40. See his *Anatome planetarium*, published in London in two parts (1675 and 1679).

41. Quoted in Dushesneau, "Malpighi," p. 113.

42. Shirley A. Roe, *Matter, Life, and Generation: Eighteenth-Century Embryology and the Haller-Wolff Debate* (Cambridge: Cambridge University Press, 1981).

43. Ibid., p. 6.

44. Ibid., pp. 21–44.

45. Ibid., p. 97.

46. See Robert Schofield, *Mechanism and Materialism: British Natural Philosophy in An Age of Reason* (Princeton: Princeton University Press, 1970), Ch. 3.

47. Ibid., p. 52.

48. Ibid., pp. 53–56.

49. Ibid., p. 69.

50. For biographical information, see G. A. Lindeboom, *Herman Boerhaave* (London: Methuen, 1968); and ibid., p. 146.

51. Quoted in Thomas S. Hall, *Ideas of Life and Matter*, vol. 1 (Chicago: University of Chicago Press, 1969), p. 369.

52. Quoted in ibid., p. 370.

53. Quoted in Schofield, *Mechanism and Materialism*, p. 193.

54. Quoted in Hall, *Ideas of Life*, p. 370.

55. Ibid., pp. 375–76.

56. See Aram Vartanian, *La Mettrie's L'Homme Machine: A Study in the Origins of an Idea* (Princeton: Princeton University Press, 1960); and Blair Campbell, "La Mettrie: The Robot and the Automaton," *Journal of the History of Ideas* 31 (1970):555–72.

57. See Hall, *Ideas of Life*, vol. 2, p. 47.

58. Vartanian, *La Mettrie*, pp. 82–89.

59. See Hall, *Ideas of Life*, vol. 2, p. 51.

60. Ibid., p. 53.

61. See Silvio Bedini, "The Role of Automata in the History of Technology," *Technology and Culture* 5 (1964):24–42; and K. Maurice and Otto Mayr, *The Clockwork Universe: German Clocks and Automata, 1550–1650* (New York: Neale Watson, 1980).

62. Chapius and Droz, *Automata*, p. 276.

63. Bedini, "Automata," p. 37.

64. Chapius and Droz, *Automata*, p. 233.

65. John Cohen, *Human Robots in Myth and Science* (South Brunswick, N.J.: A. S. Barnes, 1967), p. 87.

66. Chapius and Droz, *Automata*, p. 234.

67. Ibid., p. 293.

68. Ibid., pp. 298–99.

69. See Bedini, ''Automata,'' pp. 40–41.

70. See John W. Yolton, *Thinking Matter: Materialism in Eighteenth-Century Britain* (Minneapolis: University of Minnesota Press, 1983); and Peter Alexander, *Ideas, Qualities, and Corpuscles: Locke and Boyle on the External World* (Cambridge: Cambridge University Press, 1985).

71. Otto Mayr, *Authority, Liberty and Automatic Machinery in Early Modern Europe* (Baltimore: Johns Hopkins University Press, 1986), pp. 86–87.

72. See Corinna Delkeskamp, ''Medicine, Science, and Moral Philosophy: David Hartley's Attempt at Reconciliation,'' *The Journal of Medicine and Philosophy* 2 (1977):166–76; and Schofield, *Mechanism and Materialism*, pp. 198–99.

73. M. H. Abrams, *The Mirror and Lamp: Romantic Theory and the Critical Tradition* (Oxford: Oxford University Press, 1953), pp. 159–67.

74. See Peter Heimann, ''Molecular Forces, Statistical Representation, and Maxwell's Demon,'' *Studies in History and Philosophy of Science* 1 (1970):189–211. Heimann shows that Maxwell ruled out the possibility of reducing statistical laws to the effects of Newtonian mechanics.

Chapter 4

1. See Morris Berman, *The Reenchantment of the World* (Ithaca: Cornell University Press, 1981).

2. See Marshall Clagett, *Greek Science in Antiquity* (New York: Collier, 1955); and G. E. R. Lloyd, *Aristotle: The Growth and Structure of His Thought* (Cambridge: Cambridge University Press, 1968).

3. Stephen Toulmin and June Goodfield, *The Architecture of Matter* (Chicago: University of Chicago Press, 1962), pp. 76–77.

4. R. Hooykaas, *Religion and The Rise of Modern Science* (Grand Rapids, Mich.: William B. Eerdmans, 1972), pp. 3–5.

5. John Herman Randall, Jr., *Aristotle* (New York: Columbia University Press, 1960), pp. 220–24; and Frederick Solmsen, *Aristotle's System of the Physical World* (Ithaca: Cornell University Press, 1960).

6. Toulmin and Goodfield, *Architecture of Matter*, pp. 86–87.

7. Randall, *Aristotle*, p. 112.

8. Ibid., p. 126.

9. S. Sambursky, *The Physics of the Stoics* (London: Routledge & Kegan Paul, 1959); and Toulmin and Goodfield, *Architecture of Matter*, pp. 94–95.

10. See Frances Yates, *Giordano Bruno and the Hermetic Tradition* (Chicago: University of Chicago Press, 1979); D. P. Walker, *Spiritual and Demonic Magic: From Ficino to Campanella* (Notre Dame, Ind.: University of Notre Dame Press, 1975); and Allen Debus, *The Chemical Philosophy: Paracelsian Science and Medicine in the Sixteenth and Seventeenth Centuries*, 2 vols. (New York: Science History Publications, 1977).

11. For the origins, see Mircea Eliade, *The Forge and the Crucible*, 2d ed. (Chicago: University of Chicago Press, 1978); and Allen Debus, ''Alchemy,'' *Dictionary of the History of Ideas*, 5 vols. (New York: Scribner's Sons, 1973–74), 1:27–34.

12. See Geoffrey Chaucer, *The Canterbury Tales*, trans. Nevell Coghill (Baltimore: Penguin Books, 1960), pp. 472–77.

13. See Debus, *Chemical Philosophy*, I:4–5.

14. See Eliade, *Forge and the Crucible,* p. 33.
15. Titus Burckhardt, *Alchemy: Science of the Cosmos, Science of the Soul* (Baltimore: Penguin Books, 1971), pp. 78–81.
16. Debus, *Chemical Philosophy,* I:8.
17. See R. Hooykaas, "Chemical Trichotomy before Paracelsus?" *Archives Internationales d'Histoire des Sciences* 28 (1949):1063–74.
18. Debus, *Chemical Philosophy,* I:9–10.
19. Ibid., p. 11.
20. See W. P. D. Wightman, *Science and the Renaissance,* 2 vols. (New York: Hafner, 1962), 1:16.
21. See Frances Yates, "The Hermetic Tradition in Renaissance Science," in Charles Singleton, ed., *Art, Science and History in the Renaissance* (Baltimore: Johns Hopkins University Press, 1967), pp. 255–74.
22. Ibid., p. 256. In 1614 Isaac Casaubon correctly concluded that the *Corpus* was written in the first and second century A.D. by several authors, and, instead of being the source of Greek thought, it was the product of Platonic philosophy.
23. Ibid., pp. 256–57.
24. See Walker, *Spiritual and Demonic Magic,* pp. 3–29.
25. Ibid., p. 32.
26. Quoted in ibid., p. 13.
27. Ibid., p. 32.
28. Ibid., pp. 14–15.
29. Quoted in Allen Debus, *Man and Nature in the Renaissance* (Cambridge: Cambridge University Press, 1978), p. 82. Also see Thomas Kuhn, *The Copernican Revolution: Planetary Astronomy in the Development of Western Thought* (New York: Random House, 1959), p. 131.
30. See Yates, *Giordano Bruno,* pp. 190–397.
31. See Yates, "Hermetic Tradition," pp. 268–69; and E. Zilsel, "The Origins of William Gilbert's Scientific Method," *Journal of the History of Ideas* 2 (1941):4.
32. See Allen G. Debus, *Chemistry, Alchemy and the New Philosophy, 1550–1700* (London: Variorum Reprints, 1987); and Debus, *Chemical Philosophy.*
33. See Charles Webster, *From Paracelsus to Newton: Magic and the Making of Modern Science* (Cambridge: Cambridge University Press, 1983); and Walter Pagel, *Paracelsus: An Introduction to Philosophical Medicine in the Era of the Renaissance,* 2d rev. ed. (New York: S. Karger, 1982).
34. Debus, *Chemical Philosophy,* I:69–77.
35. Many times the Paracelsians referred to both the four elements and the three principles.
36. See Debus, *Chemical Philosophy,* pp. 84–96.
37. Quoted in ibid., p. 87.
38. Ibid., pp. 96–117.
39. Walter Pagel, *Joan Baptista van Helmont: Reformer of Science and Medicine* (Cambridge: Cambridge University Press, 1982).
40. Toulmin and Goodfield, *Architecture of Matter,* pp. 150–56.
41. See Theodore M. Brown, "From Mechanism to Vitalism in Eighteenth-Century Physiology," *Journal of the History of Biology* 7 (1974):179-216; and June Goodfield-Toulmin, "Some Aspects of English Physiology: 1780–1840," *Journal of the History of Biology* 2 (1969):283–320.
42. See Lester S. King, "Stahl and Hoffmann: A Study in Eighteenth-Century Animism," *Journal of the History of Medicine* 19 (1964):118–30; and Robert E. Schofield,

Mechanism and Materialism: British Natural Philosophy in An Age of Reason (Princeton: Princeton University Press, 1969), pp. 200–201.

43. See Goodfield-Toulmin, "Some Aspects of English Physiology," pp. 291–292.
44. See Shirley A. Roe, *Matter, Life, and Generation: Eighteenth-Century Embryology and the Haller-Wolff Debate* (Cambridge: Cambridge University Press, 1981), pp. 13–20.
45. See Brown, "From Mechanism to Vitalism," pp. 181–83.
46. See David M. Knight, "The Vital Flame," *Ambix* 23 (1976):5–15, esp. p. 7.
47. Quoted in Goodfield-Toulmin, "Some Aspects of English Physiology," p. 291.
48. See J. L. Heilbron, *Electricity in the 17th and 18th Centuries: A Study of Early Modern Physics* (Berkeley: University of California Press, 1979), pp. 491–92; Paul Fleury Mottelay, *Bibliographical History of Electricity and Magnetism* (New York: Arno Press, 1975), pp. 283–85; Bern Dibner, *Luigi Galvani* (Norwalk, Conn.: Brundy Library, 1971); and Knight, "Vital Flame."
49. Mottelay, *History of Electricity,* pp. 304–6.
50. Christopher Small, *Mary Shelley's Frankenstein: Tracing the Myth* (Pittsburgh: University of Pittsburgh Press, 1973), p. 333.
51. Knight, "Vital Flame," p. 8.
52. Goodfield-Toulmin, "Some Aspects of English Physiology," p. 299.
53. For a good survey of *naturphilosophie,* see Stephen F. Mason, *A History of the Sciences,* rev. ed. (New York: Collier, 1962), pp. 394–62. Also see H. A. M. Snelders, "Romanticism and *Naturphilosophie* and the Inorganic Natural Science 1797–1840: An Introductory Survey," *Studies in Romanticism* 9 (1970): 193–215; David Knight, *The Age of Science* (Oxford: Basil Blackwell, 1986); David Knight, "The Physical Sciences and the Romantic Movement," *History of Science* 9 (1970):54–75; and Knight, "Vital Flame," pp. 8–9.
54. Quoted in L. Pearce Williams, *Michael Faraday: A Biography* (New York: Simon and Schuster, 1971), p. 61.
55. See Frederick Copleston, *A History of Philosophy, Volume 7, Modern Philosophy, Part 1, Fichte to Hegel* (Garden City, N.Y.: Image Books, 1965), p. 127.
56. Snelders, "Romanticism," p. 197.
57. Copleston, *History of Philosophy,* p. 139.
58. Snelders, "Romanticism," pp. 197–200.
59. Williams, *Faraday,* p. 59.
60. Snelders, "Romanticism," p. 203.
61. Williams, *Faraday,* p. 140.
62. Ibid., pp. 53–95.
63. Ibid., pp. 408–64.
64. Mason, *History of Sciences,* pp. 478–79.
65. *Encyclopaedia Britannica,* 9th ed., s.v. "Lorenz Oken."
66. Mason, *History of Sciences,* p. 357.
67. Quoted in Mottelay, *History of Electricity,* p. 404.
68. See Schofield, *Mechanism and Materialism,* p. 235.
69. See Ernst Mayr, *The Growth of Biological Thought: Diversity, Evolution, and Inheritance* (Cambridge: Belknap Press of Harvard University Press, 1982), pp. 140–49.
70. See James L. Larson, *Reason and Experience: The Representation of Natural Order in the Work of Carl von Linné* (Berkeley: University of California Press, 1971); and F. A. Stafleu, *Linnaeus and the Linnaeans* (Utrecht: International Association for Plant Taxonomy, 1971).

71. Mayr, *Growth of Biological Thought,* pp. 190–208.
72. Michel Foucault, *The Order of Things: An Archaeology of the Human Sciences* (New York: Vintage, 1970).
73. Ibid., p. 226.
74. Ibid., p. 227.
75. Ibid., p. 231.
76. Ibid., p. 267.
77. See Michael Ruse, *The Darwinian Revolution; Science Red in Tooth and Claw* (Chicago: University of Chicago Press, 1979); Charles C. Gillispie, *Genesis and Geology: A Study in the Relations of Scientific Thought, Natural Theology and Social Opinion in Great Britain 1790–1850* (Cambridge: Harvard University Press, 1951); John C. Greene, *The Death of Adam: Evolution and Its Impact on Western Thought* (Ames: Iowa State University Press, 1959); and Mayr, *Growth of Biological Thought.*
78. See Ruse, *Darwinian Revolution,* pp. 343–93; and Mayr, *Growth of Biological Thought,* pp. 3–15.
79. See Peter J. Bowler, *The Non-Darwinian Revolution; Reinterpreting a Historical Myth* (Baltimore: Johns Hopkins University Press, 1988); Peter J. Bowler, *The Eclipse of Darwinism: Anti-Darwinian Evolution Theories in the Decades around 1900* (Baltimore: Johns Hopkins University Press, 1983); and David L. Hull, *Darwin and His Critics: The Reception of Darwin's Theory of Evolution by the Scientific Community* (Cambridge: Harvard University Press, 1973).
80. See Richard Burkhardt, *The Spirit of the System* (Cambridge: Harvard University Press, 1977), pp. 127–42; and Richard Burkhardt, "The Inspiration of Lamarck's Belief in Evolution," *Journal of the History of Biology* 5 (1972):413–38; Madeleine Barthélemy-Madaule, *Lamarck the Mythical Precursor: A Study of the Relations Between Science and Ideology,* trans. M. H. Shank (Cambridge, Mass.: MIT Press, 1982); Ruse, *Darwinian Revolution,* pp. 5–12; and Mayr, *Growth of Biological Thought,* pp. 343–57.
81. For a discussion of Darwin's theory of evolution, see Neal C. Gillespie, *Charles Darwin and the Problem of Creation* (Chicago: University of Chicago Press, 1979); Loren Eiseley, *Darwin's Century: Evolution and the Men Who Discovered It* (Garden City, N.Y.: Doubleday, 1958); Gertrude Himmelfarb, *Darwin and the Darwinian Revolution* (New York: Norton, 1962); Ruse, *Darwinian Revolution;* Mayr, *Growth of Biological Thought;* Bowler, *Non-Darwinian Revolution;* and Bowler, *Eclipse of Darwinism.*
82. See Mayr, *Growth of Biological Thought,* pp. 525–30; Bowler, *Non-Darwinian Revolution;* and Bowler, *Eclipse of Darwinism.*
83. See A. Hunter Dupree, *Asa Gray* (New York: Atheneum, 1968).
84. Bowler, *Non-Darwinian Revolution,* p. 84.
85. See D. C. Phillips, "Organicism in the Late Nineteenth and Early Twentieth Centuries," *Journal of the History of Ideas* 31 (1970):413–32; and F. W. Coker, *Organismic Theories of the State* (New York: AMS Press, 1967).
86. See W. M. Simon, *European Positivism in the Nineteenth Century* (Ithaca: Cornell University Press, 1963).
87. J. D. Y. Peel, *Herbert Spencer: The Evolution of a Sociologist* (New York: Basic Books, 1971).
88. Richard Hofstadter, *Social Darwinism in American Thought,* rev. ed. (Boston: Beacon Press, 1955).
89. Herbert Spencer, *First Principles,* 6th ed. (New York: D. Appleton Co., 1900), pp. 280–83.

90. Ibid., p. 367.
91. Ibid., p. 300.
92. See Anthony Giddens, *Emile Durkheim* (London: Penguin, 1978).
93. Ibid., p. 31.
94. Ibid., pp. 31–32.
95. See M. H. Abrams, *The Mirror and the Lamp: Romantic Theory and the Critical Tradition* (Oxford: Oxford University Press, 1953), pp. 170–77.

Chapter 5

1. See Karl von Frisch, *Animal Architecture* (New York: Harcourt, Brace, Jovanovich, 1974); and M. J. French, *Invention and Evolution: Design in Nature and Engineering* (Cambridge: Cambridge University Press, 1988).
2. Mircea Eliade, *The Forge and the Crucible,* 2d ed. (Chicago: University of Chicago Press, 1978), pp. 181–82.
3. Quoted in Eliade, *Forge and the Crucible,* p. 47.
4. Ibid., p. 31.
5. Ibid., pp. 71–74.
6. Ibid., p. 29.
7. Ibid., p. 57.
8. See John Herman Randall, Jr., *Aristotle* (New York: Columbia University Press, 1960), pp. 272–76.
9. Aristotle *Physics* 2.8.199a12–15, quoted in Randall, *Aristotle,* p. 87.
10. Aristotle *Physics* 2.8.199a16–18, quoted in Randall, *Aristotle,* p. 275.
11. Randall, *Aristotle,* p. 276.
12. R. Hooykaas, *Religion and the Rise of Modern Science* (Grand Rapids, Mich.: William B. Eerdmans, 1972), p. 55.
13. Friedrich Klemm, *A History of Western Technology,* trans. Dorothea Waley Singer (Cambridge, Mass.: MIT Press, 1964), p. 70.
14. See William Newman, "Technology and Alchemical Debate in the Late Middle Ages," *Isis* 80 (1989):423–45; and Lynn White, Jr., "Cultural Climates and Technological Advance in the Middle Ages," *Viator* 2 (1971):171–201.
15. Quoted in Klemm, *History of Technology,* p. 71.
16. Newman, "Technology and Alchemical Debate," p. 424.
17. Ibid.
18. William Eamon, "Technology as Magic in the Late Middle Ages and Renaissance," *Janus* 70 (1983):171–212.
19. Pierre de Maricourt, quoted in Klemm, *History of Technology,* p. 92.
20. For a discussion of the golem, see Gershom Scholem, *On the Kabbalah and Its Symbolism* (New York: Schoken, 1969), pp. 158–204.
21. Ibid., p. 199.
22. Frances Yates, "The Hermetic Tradition in Renaissance," in Charles S. Singleton, ed., *Art, Science, and History in the Renaissance* (Baltimore: Johns Hopkins University Press, 1967), p. 260.
23. Ibid., pp. 255–59.
24. Ibid., p. 260.
25. Ibid. Also see D. P. Walker, *Spiritual and Demonic Magic: From Ficino to Campanella* (Notre Dame, Ind.: University of Notre Dame Press, 1975), p. 135 n.1.
26. Yates, "Hermetic Tradition," pp. 260–61. Also see Eugenio Garin, *Science and Civic*

Life in the Italian Renaissance (Garden City, N.Y.: Doubleday and Co., 1969), p. 71.

27. Leonardo Da Vinci, *The Literary Works of Leonardo Da Vinci,* ed. J. P. Richter (Oxford: Oxford University Press, 1939), quoted in Klemm, *History of Western Technology,* p. 126.

28. Ibid., p. 127.

29. Bombastus van Holenheim (Paracelsus), *Das Buch Paragranum,* from *Paracelsus' Selected Works* (London: Routledge & Kegan Paul, 1951), quoted in Klemm, *History of Western Technology,* p. 143.

30. See John Baptista della Porta, *Natural Magick* (1658), rpt., with intro. by Derek Price (New York: Basic Books, 1957).

31. Ibid., p. 2.

32. Ibid., p. 3.

33. Ibid.

34. Frances Yates, *Theatre of the World* (Chicago: University of Chicago Press, 1969).

35. Ibid., p. 31.

36. Ibid., p. 5.

37. See John Dee, *The Mathematical Preface to the Elements of Geometrie of Euclid of Megara* (1570), with intro. by Allen Debus (New York: Scientific History Publications, 1975).

38. Ibid., p. 21.

39. Ibid., "Thaumaturgike."

40. Ibid. Also see Yates, *Theatre of the World,* p. 30.

41. Yates, *Theatre of the World,* p. 20–59.

42. Ibid., pp. 42–79.

43. Ibid., pp. 136–61; and Frances Yates, *The Art of Memory* (Chicago: University of Chicago Press, 1966).

44. Yates, *Theatre of the World,* pp. 142–43.

45. Ibid., p. 172.

46. See Frances Yates, *The Rosicrucian Enlightenment* (London: Routledge & Kegan Paul, 1972), pp. 1–14.

47. Silvio Bedini, "The Role of Automata in the History of Technology," *Technology and Culture* 5 (1964):25–28.

48. See Rhys Jenkins, *The Collected Papers of Rhys Jenkins* (Cambridge: Printed for Newcomen Society by Cambridge University Press, 1936), pp. 48–97.

49. See D. S. L. Cardwell, *From Watt to Clausius: The Rise of Thermodynamics in the Early Industrial Age* (Ithaca: Cornell University Press, 1971).

50. See Terry S. Reynolds, *Stronger than A Hundred Men: A History of the Vertical Water Wheel* (Baltimore: Johns Hopkins University Press, 1983).

51. See David M. Knight, "The Vital Flame," *Ambix* 23 (1976):5–15.

52. See E. Mendelsohn, *Heat and Life* (Cambridge: Harvard University Press, 1964); and G. J. Goodfield, *The Growth of Scientific Physiology: Physiological Method and the Mechanist–Vitalist Controversy, Illustrated by the Problem of Respiration and Animal Heat* (London: Hutchinson, 1960).

53. See Jenkins, *Collected Papers,* pp. 94–97.

54. Ibid., p. 63.

55. See Eugene S. Ferguson, "Kinematics of Mechanisms from the Time of Watt," *United States National Museum Bulletin* no. 228 (Washington, D.C.: Smithsonian Institution, 1962); Richard Hartenberg, and Jacques Denavit, *Kinematic Synthesis of Linkages* (New York: McGraw-Hill, 1964); Franz Reuleaux, *The Kinematics of*

Machinery: Outlines of a Theory of Machines, trans. Alex B. W. Kennedy (London: Macmillan, 1876), p. 9; and Brigitte Hoppe, "Biologische und technische Bewegungslehre im 19. Jahrhundert," *Georg-Agricola-Gesellschaft, Geschichte der Naturwissenschaften und der Technik im 19. Jahrhundert* (Düsseldorf: VDI-Verlag, 1969), pp. 9–35.

56. Robert Willis, *Principles of Mechanism: Designed for Students in Universities, and for Engineering Students Generally,* 2d ed. (London: Longmans, 1870), p. vii.

57. Quoted in Willis, *Principles of Mechanism,* p. viii.

58. Willis, *Principles of Mechanism,* p. xiv.

59. Reuleaux, *Kinematics,* p. 17.

60. H. H. Suplee, "Franz Reuleaux, In Memorium," *Transactions of the American Society of Mechanical Engineers* 26 (1905):816.

61. Thomas Ewbank, *A Descriptive and Historical Account of Hydraulic and Other Machines of Raising Water* (London: Tilt and Bogue, 1842), p. 258.

62. Ibid.

63. For biographical information, see Jessie Aitken Wilson, *Memoir of George Wilson* (London: Macmillan, 1862).

64. George Wilson, *What is Technology?* (Edinburgh: Sutherland and Knox, 1855); and George Wilson, "On the Physical Sciences Which Form the Basis of Technology," *The Edinburgh New Philosophical Journal* n.s. 5 (1857):64–101. Also see the brief review of *What is Technology?* in *The Edinburgh New Philosophical Journal* n.s. 3 (1856):156–58.

65. Wilson, *What is Technology?,* p. 4.

66. Ibid., p. 8.

67. Wilson, "On the Physical Sciences," pp. 64–65.

68. Ibid., p. 91.

69. Ibid., p. 93.

70. George Wilson, "On the Fruits of the Cucurbitaceae and Crescentiacea, as the Original Models of Various Clays, Glass, Metallic, and Other Hollow or Tubular Vessels, and Instruments Employed in the Arts," *The Edinburgh New Philosophical Journal* n.s. 10(1859):279–83, esp. 280.

71. George Wilson, "On the Electrical Fishes as the Earliest Electrical Machines Employed by Mankind," *The Edinburgh New Philosophical Journal* n.s. 6(1857):267–87, esp. 286.

72. Wilson, "On the Physical Sciences," p. 92.

73. Ibid., p. 93.

74. See George Basalla, *The Evolution of Technology* (Cambridge: Cambridge University Press, 1988).

75. Herbert Spencer, *First Principles,* 6th ed. (New York: D. Appleton Co., 1900), p. 297.

76. Ibid.

77. See Philip Steadman, *The Evolution of Designs: Biological Analogy in Architecture and the Applied Arts* (Cambridge: Cambridge University Press, 1979), p. 86.

78. Ibid., pp. 86–87; and Basalla, *Evolution of Technology,* pp. 17–21.

79. H. Balfour, introduction to Lt.-Gen. A. Lane-Fox Pitt-Rivers, *The Evolution of Culture and Other Essays,* ed. J. L. Myres (Oxford, 1906), p. v, quoted in Steadman, *Evolution of Designs* p. 87.

80. Steadman, *Evolution of Designs,* pp. 89–90.

81. Ibid., p. 88.

82. Ibid., p. 90.

83. See M. W. Thompson, *General Pitt-Rivers: Evolution and Archaeology in the Nineteenth Century* (Bradford-on-Avon: Moonraker Press, 1977).

84. See letter from Marx to Engels (January 28, 1863) published in *The Essential Marx: The Non-Economic Writings,* ed. and trans. Saul Padover (New York: New American Library, 1978), p. 362. For a study of Marx's life, see Isaiah Berlin, *Karl Marx: His Life and Environment* (New York: Oxford University Press, 1963).

85. Karl Marx, *Capital,* vol. 1, ed. Frederick Engels, trans. Samuel Moore and Edward Aveling (New York: International Publishers, 1977), p. 372.

86. Ibid., pp. 179–80.

87. Ibid., p. 180.

88. Ibid., pp. 381–82.

89. Andrew Ure quoted in ibid., pp. 418–19.

90. Marx, *Capital,* p. 419.

91. Ibid., p. 420.

92. See Steadman, *Evolution of Designs,* pp. 28–33.

93. Quoted in ibid., pp. 41–42.

94. Ibid., pp. 29–30.

95. See Peter Collins, *Changing Ideals in Modern Architecture, 1750–1950* (London: Faber and Faber, 1965), p. 155.

96. See Fred Majdalany, *The Eddystone Light* (Boston: Houghton Mifflin Co., 1960), pp. 113–21.

97. See Folke T. Kihlstedt, "The Crystal Palace," *Scientific American* 251 (October, 1984):132–43.

98. See Carl W. Condit, "Sullivan's Skyscrapers and the Expressions of Nineteenth-Century Technology," *Technology and Culture* 1 (1959):78–93; and Narciso G. Menocal, *Architecture as Nature: The Transcendentalist Idea of Louis Sullivan* (Madison: University of Wisconsin Press, 1981).

99. Quoted in Collins, *Changing Ideals,* p. 151.

100. See Herbert Sussman, *Victorians and the Machine: The Literary Response to Technology* (Cambridge: Harvard University Press, 1968).

Chapter 6

1. See Timothy Lenoir, *The Strategy of Life: Teleology and Mechanics in Nineteenth-Century Biology* (Dordrecht: D. Reidel, 1982), p. 238.

2. Samuel Butler, *Erewhon or Over the Range,* ed. Hans-Peter Breuer and Daniel F. Howard (Newark: University of Delaware Press, 1981).

3. Philip Steadman, *The Evolution of Designs: Biological Analogy in Architecture and the Applied Arts* (Cambridge: Cambridge University Press, 1979), pp. 130–31.

4. Introduction to Butler's *Erewhon,* p. 18.

5. Herbert Sussman, *Victorians and the Machine: The Literary Response to Technology* (Cambridge: Harvard University Press, 1968), pp. 151–52.

6. Butler, *Erewhon,* pp. 184–85.

7. Ibid., p. 189.

8. Ibid., p. 190.

9. Ibid., p. 203.

10. Ibid., pp. 203–4.

11. Ibid., p. 203.

12. Sussman, *Victorians and the Machine,* p. 154.

13. See Hilde Hein, "The Endurance of the Mechanism-Vitalism Controversy," *Journal of the History of Biology* 5 (1972):159–88.
14. See Lenoir, *Strategy of Life.*
15. Ibid., p. 14.
16. Stephen Toulmin and June Goodfield, *Architecture of Matter* (Chicago: University of Chicago Press, 1962), p. 326.
17. William Coleman, *Biology in the Nineteenth Century: Problems of Form, Function, and Transformation* (Cambridge: Cambridge University Press, 1977), pp. 146–47.
18. Thomas S. Hall, *Ideas of Life and Matter,* vol. 2 (Chicago: University of Chicago Press, 1969), pp. 258–63.
19. Quoted in ibid., p. 269. Also see Timothy O. Lipman, "Vitalism and Reductionism in Leibig's Physiological Thought," *Isis* 58 (1967):170–73.
20. Hall, *Life and Matter,* 2:271.
21. See Lenoir, *Strategy of Life,* Ch. 3.
22. Toulmin and Goodfield, *Architecture of Matter,* pp. 339–43.
23. Coleman, *Biology in the Nineteenth Century,* pp. 19–20.
24. Hall, *Life and Matter,* 2:179.
25. Quoted in Toulmin and Goodfield, *Architecture of Matter,* p. 347.
26. See Gerd Buchdahl, "The Leading Principles of Induction: The Methodology of Matthias Schleiden," in *Foundations of Scientific Method: The Nineteenth Century,* Ronald M. Giere and Richard S. Westfall, eds. (Bloomington: Indiana University Press, 1973), pp. 23–52; and Lenoir, *Strategy of Life,* pp. 124–34.
27. Quoted in Hall, *Life and Matter,* 2:188.
28. Coleman, *Biology in the Nineteenth Century,* p. 26.
29. Quoted in Toulmin and Goodfield, *Architecture of Matter,* p. 349.
30. See Hall, *Life and Matter,* 2:215.
31. See Gerald Geison, "The Protoplasmic Theory of Life and the Vitalist–Mechanist Debate," *Isis* 60 (1969):273–92.
32. Hall, *Life and Matter,* 2:280.
33. Ibid.
34. Ibid., p. 283.
35. Quoted in ibid., pp. 283–84.
36. See Lenoir, *Strategy of Life,* Ch. 5.
37. Quoted in Hall, *Life and Matter,* 2:278.
38. Frederic L. Holmes, *Claude Bernard and Animal Chemistry* (Cambridge: Harvard University Press, 1974); and J. M. D. Olmsted and E. Harris Olmsted, *Claude Bernard and the Experimental Method in Medicine* (New York: Henry Schuman, 1952).
39. Quoted in Hall, *Life and Matter,* 2:301.
40. Claude Bernard, "Phenomenes de la vie," quoted in Claude Bernard, *An Introduction to the Study of Experimental Medicine,* intro. Lawrence J. Henderson (New York: Dover, 1957), p. vii.
41. Coleman, *Biology in the Nineteenth Century,* pp. 157–58.
42. Hall, *Life and Matter,* 2:303.
43. Quoted in ibid., p. 302.
44. Everett Mendelsohn, "Physical Models and Physiological Explanation in Nineteenth-Century Biology," *The British Journal for the History of Science* 2 (1965):201–19.
45. See Hein, "Mechanism–Vitalism Controversy," pp. 159–88.
46. See Garland Allen, *Life Science in the Twentieth Century* (Cambridge: Cambridge University Press, 1978), pp. 29–34.
47. Ibid., pp. 34–35.

48. For a study of Loeb, see Philip J. Pauly, *Controlling Life: Jacques Loeb and the Engineering Ideal in Biology* (New York: Oxford University Press, 1987).
49. Quoted in ibid., p. 51.
50. See ibid., pp. 100–105.
51. See ibid., pp. 80–81; and Allen, *Life Science,* pp. 130–63.
52. See Allen, *Life Science,* p. 32.
53. Ibid., pp. 100–103.
54. See Donna Haraway, *Crystals, Fabrics, and Fields: Metaphors of Organicism in Twentieth-Century Developmental Biology* (New Haven: Yale University Press, 1976), p. 115; Viktor Hamburger, "Hans Spemann and the Organizer Concept," *Experientia* 25 (1969):1121–25; and Allen, *Life Science,* pp. 114–22.
55. See Jane Oppenheimer, "Cells and Organizers," *American Zoologist* 10 (1970):75–88.
56. See Haraway, *Crystals, Fabrics, and Fields,* Ch. 5.
57. Ibid., Ch. 2; and Allen, *Life Science,* pp. 104–6.
58. Lawrence J. Henderson, *Order of Nature* (Cambridge: Harvard University Press, 1917).
59. J. S. Haldane, *Organism and Environment* (New Haven: Yale University Press, 1917).
60. W. E. Ritter, *The Unity of the Organism* (Boston: Gorham Press, 1919); Ludwig von Bertalanffy, *Kritisch Theorie der Formbildung* (Berlin: Borntraeger, 1928); and Ludwig von Bertalanffy, *Problems of Life* (London: C. A. Watts, 1952).
61. See Robert Schofield, *Mechanism and Materialism: British Natural Philosophy in an Age of Reason* (Princeton: Princeton University Press, 1969).
62. See Jed Z. Buchwald, *The Rise of the Wave Theory of Light: Optical Theory and Experiment in the Early Nineteenth Century* (Chicago: University of Chicago Press, 1989).
63. Toulmin and Goodfield, *Architecture of Matter,* p. 251.
64. See G. N. Cantor and M. J. S. Hodge, eds., *Conceptions of Ether: Studies in the History of Ether Theories, 1740–1900* (Cambridge: Cambridge University Press, 1981).
65. See L. Pearce Williams, *Michael Faraday: A Biography* (New York: Simon and Schuster, 1971).
66. See Mary B. Hesse, *Forces and Fields: A Study of Action at a Distance in the History of Physics* (Totowa, N.J.: Littlefield, Adams, 1965), p. 200; and Charles C. Gillespie, *The Edge of Objectivity* (Princeton: Princeton University Press, 1960), pp. 456–57.
67. Williams, *Faraday,* p. 380.
68. See John Hendry, *James Clerk Maxwell and the Theory of the Electromagnetic Field* (Boston: Adams Hilger, 1986). Hendry argues that Maxwell was not a traditional mechanist.
69. Gillispie, *Edge of Objectivity,* pp. 468–69.
70. Quoted in Toulmin and Goodfield, *Architecture of Matter,* p. 258.
71. See Gillispie, *Edge of Objectivity,* p. 473.
72. *Dictionary of Scientific Biography,* s.v. "H. A. Lorentz."
73. See P. M. Harman, *Energy, Force, and Matter: The Conceptual Development of Nineteenth-Century Physics* (Cambridge: Cambridge University Press, 1982), Ch. 4; and B. G. Doran, "Origins and Consolidation of Field Theory in Nineteenth-Century Britain: From the Mechanical to the Electrodynamic View of Nature," *Historical Studies in the Physical Sciences* 6 (1975):133–260.
74. See Martin J. Klein, "Mechanical Explanations at the End of the Nineteenth Century," *Centaurus* 17 (1972):58–82.

75. Joseph Larmor, *Aether and Matter* (Cambridge: Cambridge University Press, 1900).

76. See Russell McCormmach, "H. A. Lorentz and the Electromagnetic View of Nature," *Isis* 61 (1970):459–97; and T. Hirosige, "Origins of Lorentz's Theory of Electrons and the Concept of the Electromagnetic Field," *Historical Studies in the Physical Sciences* 1 (1969):151–209.

77. For a discussion of relativity, see Albert Einstein, "Zur Elektrodynamik bewegter Körper," *Annalen der Physik* 17 (1905):891–921; Abraham Pais, *'Subtle is the Lord . . .' The Science and Life of Albert Einstein* (Oxford: Oxford University Press, 1982); Gerald Holton, *Thematic Origins of Scientific Thought: Kepler to Einstein* (Cambridge: Harvard University Press, 1973), pp. 165–380; and Arthur I. Miller, *Albert Einstein's Special Theory of Relativity: Emergence (1905) and Early Interpretation (1905–11)* (Reading, Mass.: Addison-Wesley, 1981).

78. Mary Jo Nye, "The Nineteenth-Century Atomic Debates," *Studies in History and Philosophy of Science* 7 (1976):245–68.

79. Gillispie, *Edge of Objectivity*, p. 497.

80. Thomas Kuhn, *Black-Body Theory and the Quantum Discontinuity, 1894–1912* (Chicago: University of Chicago Press, 1978).

81. See Heinz Pagels, *The Cosmic Code: Quantum Physics and the Language of Nature* (New York: Bantam, 1982); Max Jammer, *The Philosophy of Quantum Mechanics* (New York: John Wiley and Sons, 1974); and Ruth Moore, *Niels Bohr: The Man, His Science, and the World They Changed* (Cambridge, Mass.: MIT Press, 1985).

82. David Wilson, *Rutherford: Simple Genius* (Cambridge, Mass.: MIT Press, 1983).

83. Pagels, *Cosmic Code*, pp. 56–66; and Jammer, *Philosophy of Quantum Mechanics*, pp. 20–55.

84. John Hendry, *The Creation of Quantum Mechanics and the Bohr-Pauli Dialogue* (Dordrecht: D. Reidel, 1984); Pagels, *Cosmic Code*, pp. 67–81; and Jammer, *Philosophy of Quantum Mechanics*, pp. 55–108.

85. Pagels, *Cosmic Code*, p. 75.

86. David Bohm, *Wholeness and the Implicate Order* (London: Routledge & Kegan Paul, 1980), p. 11.

87. Ibid., pp. 147–50.

88. B. J. Hiley and F. David Peat, eds., *Quantum Implications: Essays in Honour of David Bohm* (London: Routledge & Kegan Paul, 1987); and J. S. Bell, *Speakable and Unspeakable in Quantum Mechanics* (Cambridge: Cambridge University Press, 1987).

89. Bohm, *Wholeness and Implicate Order*, p. 194.

90. For example, see Max Delbrück, *Mind from Matter? An Essay on Evolutionary Epistemology*, ed. Gunther S. Stent, Ernst Peter Fischer, Solomon W. Golomb, David Presti, and Hansjakob Seiler (Oxford: Blackwell Scientific Publications, 1986). Delbrück argues the mind is less psychic and matter less materialistic after quantum mechanics.

91. Donna Haraway makes a similar suggestion for biology. See Haraway, *Crystals, Fabrics, and Fields*, Ch. 6.

92. See ibid., pp. 15, 24–25, 137–38, 184; Allen, *Life Science*, pp. 103–4; and Bohm, *Wholeness and Implicate Order*, pp. 48, 207.

93. Alfred North Whitehead, *Science and the Modern World* (New York: Free Press, 1925), p. 103.

94. Ibid., p. 72.

95. Alfred North Whitehead, *Process and Reality*, corrected ed., ed. David Ray Griffin

and Donald W. Sherburne (New York: Free Press, 1978), p. 208; and Alfred North Whitehead, *Modes of Thought* (New York: Free Press, 1938), p. 88.

96. Whitehead, *Process and Reality*, p. 208; and Whitehead, *Modes of Thought*, p. 88.

97. Whitehead, *Modes of Thought*, p. 54.

98. Ibid., p. 55.

99. Whitehead, *Process and Reality*, p. 45.

100. See Ludwig von Bertalanffy, *General System Theory*, rev. ed. (New York: George Braziller, 1968); Ludwig von Bertalanffy, *Robots, Men and Minds* (New York: George Braziller, 1967); and Bertalanffy, *Problems of Life*.

101. See Ervin Laszlo, *Introduction to Systems Philosophy: Toward a New Paradigm of Contemporary Thought* (New York: Gordon and Breach, 1972); Kenneth Boulding, *The Organizational Revolution: A Study in the Ethics of Economic Organization* (New York: Harper & Bros., 1953); and Kenneth Boulding, *Ecodynamics: A New Theory of Societal Evolution* (Beverly Hills: Sage Publications, 1978).

102. Bertalanffy, *General System Theory*, pp. 55–56, 91.

103. Herbert A. Simon, *The Sciences of the Artificial* (Cambridge, Mass.: MIT Press, 1969), Ch. 4.

104. Ibid., pp. 99–100.

105. For a discussion of the anthropic principle, see John D. Barrow and Frank J. Tipler, *The Anthropic Cosmological Principle* (Oxford: Oxford University Press, 1988); George Gale, "The Anthropic Principle," *Scientific American* 245 (December, 1981):154–71; and Stephen W. Hawking, *A Brief History of Time: From the Big Bang to Black Holes* (New York: Bantam, 1988). For a discussion of the Gaia hypothesis, see James E. Lovelock, *Gaia: A New Look at Life on Earth* (Oxford: Oxford University Press, 1982); James Lovelock, *The Ages of Gaia: A Biography of Our Living Earth* (New York: Bantam, 1990); and Dorion Sagan and Lynn Margulis, "Gaia and Philosophy," in *On Nature*, ed. Leroy S. Rouner (Notre Dame, Ind.: University of Notre Dame Press, 1984), pp. 60–75.

106. Barrow and Tipler, *Anthropic Principle*, p. 16.

107. Gale, "Anthropic Principle," p. 168.

108. Barrow and Tipler, *Anthropic Principle*, pp. 21–22.

109. Quoted in ibid., p. 16.

110. Ibid., p. 4.

111. Sagan and Margulis, "Gaia and Philosophy," p. 68.

112. See Andrew J. Watson and James E. Lovelock, "Biological Homeostasis of the Global Environment: The Parable of 'Daisy World,'" *Tellus* 35b (1983):284–89; and James E. Lovelock, "Daisy World: A Cybernetic Proof of the Gaia Hypothesis," *CoEvolution Quarterly* 31 (1983):66–72.

113. Sagan and Margulis, "Gaia and Philosophy," p. 73.

Chapter 7

1. Bruce Mazlish, "The Fourth Discontinuity," in Melvin Kranzberg and William H. Davenport (eds.), *Technology and Culture: An Anthology* (New York: New American Library, 1975).

2. Ibid., p. 218.

3. See Thomas P. Hughes, *American Genesis: A Century of Invention and Technological Enthusiasm, 1870–1970* (New York: Viking, 1989), Ch. 5.

4. See Siegfried Giedion, *Mechanization Takes Command* (New York: W. W. Norton,

1948), pp. 77–127; and David Hounshell, *From the American System to Mass Production, 1800–1932* (Baltimore: Johns Hopkins University Press, 1984).

5. Giedion, *Mechanization Takes Command,* p. 84.
6. Ibid., pp. 86–93.
7. Ibid., p. 94.
8. See Hugh G. J. Aitken, *Taylorism at Watertown Arsenal: Scientific Management in Action, 1908–1915* (Cambridge, Mass.: Harvard University Press, 1960); Robert Guest, "The Rationalization of Management," in Melvin Kranzberg and Carroll Pursell, eds., *Technology and Western Civilization, Vol. 2* (New York: Oxford University Press, 1967); Daniel Nelson, *Frederick W. Taylor and the Rise of Scientific Management* (Madison: University of Wisconsin Press, 1980); and Frederick Winslow Taylor, *The Principles of Scientific Management* (New York: Norton, 1967, originally published 1911).
9. Taylor, *Scientific Management,* p. 117.
10. Ibid., p. 79.
11. Giedion, *Mechanization Takes Command,* pp. 102–6.
12. Taylor, *Scientific Management,* p. 129.
13. Ibid., pp. 140–41.
14. Ibid., p. 7.
15. See John B. Rae, *The American Automobile* (Chicago: University of Chicago Press, 1965); and John B. Rae, "The Rationalization of Production," in Kranzberg and Pursell, *Technology and Western Civilization, Vol. 2.*
16. See F. Klemm, *A History of Western Technology* (Cambridge: MIT Press, 1964), p. 333.
17. Giedion, *Mechanization Takes Command,* p. 115.
18. See ibid., pp. 125–26; and Gerald Mast, *A Short History of the Movies* (New York: Pegasus, 1971), pp. 281–82.
19. For a history of the computer, see Herman Goldstine, *The Computer: From Pascal to von Neumann* (Princeton: Princeton University Press, 1972); Pamela McCorduck, *Machines Who Think* (San Francisco: W. H. Freeman, 1979); Paul Ceruzzi, *Reckoners: The Prehistory of the Digital Computer* (Westport, Conn.: Greenwood Press, 1981); Michael R. Williams, *A History of Computing Technology* (Englewood Cliffs, N.J.: Prentice-Hall, 1985); and Joel Shurkin, *Engines of the Mind: A History of the Computer* (New York: W. W. Norton, 1984).
20. Goldstine, *The Computer,* p. 9.
21. For information on Babbage, see Anthony Hyman, *Charles Babbage: Pioneer of the Computer* (Princeton: Princeton University Press, 1982); and Henry Babbage, *Babbage's Calculating Machines,* intro. by Allan G. Bromley (Cambridge, Mass.: MIT Press, 1984).
22. For a biography of Ada Lovelace, see Dorothy Stein, *Ada: A Life and a Legacy* (Cambridge, Mass.: MIT Press, 1987).
23. See Desmond MacHale, *George Boole: His Life and Work* (Dublin: Boole Press, 1985).
24. Quoted in Goldstine, *The Computer,* p. 36.
25. Ibid., pp. 65–71.
26. Ibid., pp. 73–83.
27. See Clark R. Mollenhoff, *Atanasoff: Forgotten Father of the Computer* (Ames: Iowa State University Press, 1988); and Alice R. Burks and Arthur W. Burks, *The First Electronic Computer: The Atanasoff Story* (Ann Arbor: University of Michigan Press, 1988).

28. See Nancy Stern, *From ENIAC to UNIVAC* (Bedford, Mass.: Digital Press, 1981); and Paul Ceruzzi, "Electronics Technology and Computer Science, 1940–1975: A Coevolution," *Annals of the History of Computing* 10 (1989):257–75.

29. See McCorduck, *Machines Who Think;* J. David Bolter, *Turing's Man: Western Culture in the Computer Age* (Chapel Hill: University of North Carolina Press, 1984); Douglas Hofstadter, *Godel, Esher, Bach* (New York: Basic Books, 1979); and Frank Rose, *Into the Heart of the Mind: An American Quest for Artificial Intelligence* (New York: Harper and Row, 1984).

30. See Bolter, *Turing's Man,* pp. 43–47.

31. See McCorduck, *Machines Who Think,* pp. 40–41; and Rose, *Heart of the Mind,* p. 32.

32. See Goldstine, *The Computer,* pp. 184–203.

33. See McCorduck, *Machines Who Think,* pp. 62–69.

34. For a description of the von Neumann machine, see Bolter, *Turing's Man,* pp. 47–52.

35. Quoted in Goldstine, *The Computer,* p. 194.

36. Ibid., pp. 196–97.

37. Norbert Wiener, *Cybernetics: Or Control and Communication in the Animal and the Machine,* 2d ed. (Cambridge: MIT Press, 1961), p. 1.

38. For a history of the problem of control, see James R. Beniger, *The Control Revolution: Technological and Economic Origins of the Information Society* (Cambridge: Harvard University Press, 1986).

39. For a history of feedback, see Otto Mayr, *The Origins of Feedback Control* (Cambridge, Mass.: MIT Press, 1970).

40. Wiener, *Cybernetics,* p. 8.

41. Ibid., p. 11.

42. For a history and discussion of artificial intelligence, see Vernon Pratt, *Thinking Machines: The Evolution of Artificial Intelligence* (Oxford: Basil Blackwell, 1987); and Margaret Boden, *Artificial Intelligence and Natural Man* (New York: Basic Books, 1977).

43. Alan Turing, "Computing Machinery and Intelligence," *Mind,* vol. 59, no. 236, reprinted in James R. Newman, ed., *The World of Mathematics,* vol. 4 (New York: Simon and Schuster, 1956), pp. 2099–123, esp. p. 2099.

44. See McCorduck, *Machines Who Think,* pp. 93–114.

45. See Rose, *Heart of the Machine,* pp. 41–116.

46. McCorduck, *Machines Who Think,* pp. 252–56.

47. See Hubert L. Dreyfus and Stuart E. Dreyfus, "Making a Mind Versus Modeling the Brain: Artificial Intelligence at a Branchpoint"; and Sherry Turkle, "Artificial Intelligence and Psychoanalysis: A New Alliance," in Stephen R. Graubard, ed., *The Artificial Intelligence Debate: False Starts, Real Foundations* (Cambridge, Mass.: MIT Press, 1988), pp. 15–44, 241–68.

48. See ibid. As an example, see Daniel L. Alkon, "Memory Storage and Neural Systems," *Scientific American* 261 (July, 1989):42–50.

49. W. S. McCulloch and Walter Pitts, "A Logical Calculus of the Ideas Immanent in Neural Nets," *Bulletin of Mathematical Biophysics* 5 (1943):115–37; and D. O. Hebb, *The Organization of Behavior* (New York: Wiley, 1949).

50. Frank Rosenblatt, *Principles of Neurodynamics: Perceptrons and the Theory of Brain Mechanisms* (Washington, D.C.: Spartan Books, 1962).

51. Quoted in Dreyfus and Dreyfus, "Mind Versus Brain," pp. 19–20.

52. See Turkle, "Artificial Intelligence," p. 247.

53. See David E. Rumelhart, James L. McClelland, et al., *Parallel Distributed Process-*

ing: Explorations in the Microstructure of Cognition, 2 vols. (Cambridge, Mass.: MIT Press, 1986).

54. See Dreyfus and Dreyfus, "Mind Versus Brain," p. 16; and Turkle, "Artificial Intelligence," p. 248.

55. See Marvin L. Minsky and Seymour Papert, *Perceptrons: An Introduction to Computational Geometry,* Expanded ed. (Cambridge, Mass.: MIT Press, 1988).

56. See W. Daniel Hillis, *The Connection Machine* (Cambridge, Mass.: MIT Press, 1985); and W. Daniel Hillis, "The Connection Machine," *Scientific American* 256 (July, 1987):108–15.

57. Marvin L. Minsky, *The Society of Mind* (New York: Simon & Schuster, 1985), p. 319.

58. Minsky and Papert, *Perceptrons,* p. 269.

59. For a supporter of this position, see Hans Moravec, *Mind Children: The Future of Robot and Human Intelligence* (Cambridge: Harvard University Press, 1988).

60. See John R. Searle, "Minds, Brains, and Programs," reprinted in Douglas Hofstadter and Daniel C. Dennett, *The Mind's I* (New York: Basic Books, 1981), pp. 353–73.

61. See Roger Penrose, *The Emperor's New Mind: Concerning Computers, Minds, and the Laws of Physics* (Oxford: Oxford University Press, 1989); and Hubert Dreyfus, *What Computers Can't Do: A Critique of Artificial Reason* (New York: Harper and Row, 1972).

62. McCorduck, *Machines Who Think,* p. 181.

63. John G. Kemeny, *Man and the Computer* (New York: Charles Scribner's Sons, 1972), pp. 10–13.

64. Ibid., p. 10.

65. Ibid., p. 3.

66. See George Murdoch and John Hughes, "Clinical and Biomedical Aspects of Current Prosthetic Practice," in R. M. Kenedi, ed., *Perspectives in Biomedical Engineering* (Baltimore: University Park Press, 1973), p. 67.

67. Jasia Reichardt, *Robots* (New York: Penguin Books, 1987), p. 122.

68. Murdoch and Hughes, "Current Prosthetic Practice," p. 67.

69. Reichardt, *Robots,* p. 122.

70. D. S. Halacy, Jr., *Cyborg—Evolution of the Superman* (New York: Harper and Row, 1965), p. 25.

71. Ibid., pp. 69–70, 107–9; and Robert W. Mann, "Trade-offs at the Man-Machine Interface in Cybernetic Prostheses/Orthoses," in Kenedi, *Biomedical Engineering,* pp. 74–76.

72. For a history of hemodialysis, see J. W. Czaczkes and A. Kaplan De-Nour, *Chronic Hemodialysis as a Way of Life* (New York: Brunner/Mazel, 1978), pp. 19–44.

73. Ibid., p. 134.

74. Audrey B. Davis, *Medicine and its Technology* (Westport, Conn.: Greenwood Press, 1981), p. 11.

75. Christopher Hallowell, "Charles Lindberg's Artificial Heart," *American Heritage of Invention and Technology* 1 (Fall, 1985):58–62.

76. Halacy, *Cyborg,* p. 79.

77. See National Science Foundation, *Interactions of Science and Technology in the Innovative Process: Some Case Studies* (Washington, D.C.: National Science Foundation, Contract NSF-C 667, March 19, 1973), Ch. 5.

78. Ibid.

79. Halacy, *Cyborg,* p. 80.

80. M. M. Black, "Development and Testing of Prosthetic Heart Valves: Cardiovascular Simulation and Life Support Systems," in Kenedi, *Biomedical Engineering,* p. 21.

178 / *Notes*

81. Yukihiko Nose and W. J. Kolff, "Cardiac Assist Devices, Total Artificial Heart and Unconventional Dialysis," in Kenedi, *Biomedical Engineering,* p. 15.
82. See Gregory E. Pence, *Classic Cases in Medical Ethics* (New York: McGraw-Hill, 1990), pp. 225–50.
83. Margery Shaw, ed., *After Barney Clark* (Austin: University of Texas Press, 1984).
84. See I. L. Paul, E. L. Radin, and R. M. Rose, "Biomechanical Aspects of Orthopaedic Implants," in Kenedi, *Biomedical Engineering,* p. 95.
85. See Manfried Clynes, "Forward," in Halacy, *Cyborg,* pp. 6–8.
86. Manfried Clynes and Nathan S. Kline, "Cyborgs and Space," *Astronautics* (September, 1960), p. 27, quoted in ibid., p. 9.
87. William Coleman, *Biology in the Nineteenth Century* (Cambridge: Cambridge University Press, 1977), p. 32.
88. Stephen Toulmin and June Goodfield, *The Architecture of Matter* (Chicago: University of Chicago Press, 1962), p. 360.
89. For an early history of work on DNA, see Jeremy Cherfas, *Man-Made Life* (New York: Pantheon, 1982); and James D. Watson and John Tooze, *The DNA Story: A Documentary History of Gene Cloning* (San Francisco: W. H. Freeman, 1981).
90. Watson and Tooze, *The DNA Story,* p. 536; and Cherfas, *Man-Made Life,* p. 6.
91. Cherfas, *Man-Made Life,* p. 7.
92. See Robert Olby, "Schrödinger's Problem: What is Life?" *Journal of the History of Biology* 4 (1971):119–48.
93. Erwin Schrödinger, *What Is Life? The Physical Aspects of the Living Cell* (Cambridge: Cambridge University Press, 1944), pp. 60–62.
94. See James D. Watson, *The Double Helix* (New York: Penguin, 1970); Francis Crick, *What Mad Pursuit: A Personal View of Scientific Discovery* (New York: Basic Books, 1988); and Horace Freeland Judson, *The Eighth Day of Creation: Makers of the Revolution in Biology* (New York: Simon & Schuster, 1979).
95. Watson and Tooze, *The DNA Story,* p. 540.
96. Stephen S. Hall, *The Race to Synthesize a Human Gene* (New York: Atlantic Monthly Press, 1987).
97. See Watson, *The Double Helix,* p. 154.
98. Watson and Tooze, *The DNA Story,* pp. 543, 546.
99. François Jacob, *The Logic of Life: A History of Heredity,* trans. by Betty E. Spillmann (New York: Pantheon, 1974), p. 9.
100. Cherfas, *Man-Made Life,* p. 64.
101. For a complete test of the Supreme Court decision, see Watson and Tooze, *The DNA Story,* pp. 501–10.
102. Ibid., p. 502.
103. Ibid., p. 504.
104. Ibid., p. 505.
105. Ibid., p. 506.
106. See Wiener, *Cybernetics,* pp. 177–80.
107. Ibid., p. 180.
108. David Baltimore, "The Brain of a Cell," *Science 84* 5 (November, 1984):149–51, esp. p. 150.
109. See Moravec, *Mind Children,* p. 1; and Grant Fjermedal, *The Tomorrow Makers: A Brave New World of Living-Brain Machines* (New York: Macmillan, 1986).
110. See Heinz R. Pagels, *The Dreams of Reason: The Computer and the Rise of the Sciences of Complexity* (New York: Simon & Schuster, 1988), pp. 97–107.
111. Richard Dawkins, *The Blind Watchmaker* (New York: W. W. Norton, 1986), Ch. 3.

112. See Philip Fites, Peter Johnston, and Martin Kratz, *The Computer Virus Crisis* (New York: Van Nostrand Reinhold, 1989); A. K. Dewdeny, "Computer Recreations," *Scientific American* 250 (May, 1984):14–17; 252 (March, 1985):14–23; 256 (January, 1987):14–18; and Moravec, *Mind Children,* pp. 126–31.

113. See Fred Cohen, *Computer Viruses,* 7th DOD/NBS Computer Security Conference (Los Angeles: University of Southern California, 1984).

114. Susan Sontag, *AIDS and Its Metaphors* (New York: Farrar, Straus, and Giroux, 1988), p. 70.

115. See Fites, Johnston, and Kratz, *Computer Virus,* p. ix.

116. Ibid., pp. 133–45.

117. John Markoff, "A Darwinian Creation of Software," *The New York Times,* February 28, 1990, Sec. C, p. 6.

118. Quoted in John Markoff, "Beyond Artificial Intelligence, a Search for Artificial Life," *The New York Times,* February 25, 1990, Sec. E, p. 5.

119. Natalie Angier, "The Organic Computer," *Discover* 3 (May, 1982):76–79.

120. For a discussion of nanotechnology, see K. Eric Drexler, *Engines of Creation: The Coming Era of Nanotechnology,* Foreword by Marvin Minsky (New York: Anchor Books, 1987); and A. K. Dewdeny, "Nanotechnology: Wherein Molecular Computers Control Tiny Circulatory Submarines," *Scientific American* 358 (January, 1988):100–103.

121. K. Eric Drexler, "Molecular Engineering: An Approach to the Development of General Capabilities for Molecular Manipulation," in *Proceedings of the National Academy of Sciences of the United States of America* 78 (September, 1981):5275–78.

122. See Fjermedal, *Tomorrow Makers,* pp. 245–47.

123. See Thomas Pynchon, "Is It O.K. to Be a Luddite?" *New York Review of Books* (October 28, 1984), p. 41.

Chapter 8

1. See A. Pablo Iannone, ed., *Contemporary Moral Controversies in Technology* (New York: Oxford University Press, 1987); and David Suzuki and Peter Knudtson, *Genethics: The Clash between the New Genetics and Human Values* (Cambridge: Harvard University Press, 1989).

2. Hans Jonas, *Philosophical Essays: From Ancient Creed to Technological Man* (Englewood Cliffs, N.J.: Prentice-Hall, 1974), pp. 3–7.

3. Ibid., p. 6.

4. Ibid., p. 8.

5. See Bernard Gendron, "The Viability of Environmental Ethics," in Paul Durbin and Friedrich Rapp, eds., *Philosophy and Technology* (Dordrecht: D. Reidel, 1983), p. 191.

6. See Herbert Schadelbach, "Is Technology Ethically Neutral?" in Melvin Kranzberg, ed., *Ethics in an Age of Pervasive Technology* (Boulder: Westview Press, 1980), pp. 28–30.

7. For a discussion of exploitation, see Gendron, "Environmental Ethics," pp. 187–88.

8. For a discussion of the new attitudes toward nature, see Keith Thomas, *Man and the Natural World* (New York: Pantheon, 1983), pp. 173–81.

9. See Peter Singer, *Animal Liberation* (New York: A New York Review Book, 1975).

10. See Jacques Ellul, *The Technological Society,* trans. John Wilkinson (New York:

Alfred A. Knopf, 1964); and Langdon Winner, *Autonomous Technology: Technics-out-of-Control as a Theme in Political Thought* (Cambridge, Mass.: MIT Press, 1977.

11. For a discussion of the relationship between animism and autonomous technology, see Winner, *Autonomous Technology,* pp. 30–43.

12. See Karel Capek, *R.U.R.,* reprinted in Arthur O. Lewis, Jr., ed., *Of Men and Machines* (New York: E. P. Dutton, 1963), pp. 3–58.

13. Ibid., p. 17.

14. Ibid., p. 31.

15. Ibid., p. 44.

16. Ibid., p. 55.

17. Ibid., pp. 16–17.

18. Ibid., p. 31.

19. Satosi Watanabe, "Comments on Key Issues," in Sidney Hook, ed., *Dimensions of Mind: A Symposium* (New York: New York University Press, 1960), p. 146.

20. Hilary Putnam, "Minds and Machines," in Hook, *Dimensions of Mind,* pp. 148–79.

21. Ibid., p. 150.

22. Ibid., pp. 162–66.

23. Hilary Putnam, "Robots: Machines or Artificially Created Life?" *The Journal of Philosophy* 61 (1964):668-91, esp. p. 677.

24. Ibid., p. 678.

25. Alan Turing, "Computing Machinery and Intelligence," reprinted in Alan Ross Anderson, ed., *Minds and Machines* (Englewood Cliffs, N.J.: Prentice-Hall, 1964), pp. 20–21.

26. Paul Ziff, "The Feelings of Robots," reprinted in Anderson, *Minds and Machines,* pp. 98–103.

27. Putnam, "Robots: Machines or Artificially Created Life?", p. 680.

28. Turing, "Computing Machinery and Intelligence," p. 21.

29. See J. J. C. Smart, "Professor Ziff on Robots," in Anderson, *Minds and Machines,* pp. 104–5; and Ninian Smart, "Robots Incorporated," in Anderson, *Minds and Machines,* pp. 106–8; and Michael Scriven, "The Complete Robot: A Prolegomena to Andriodology," in Hook, *Dimensions of Mind,* pp. 121–24.

30. Norbert Wiener, "The Brain and the Machine," in Hook, *Dimensions of Mind,* pp. 113–17.

31. Hubert Dreyfus, *What Computers Can't Do: A Critique of Artificial Reason* (New York: Harper and Row, 1972), pp. xv–xvii.

32. Ibid., pp. 203–6.

33. See Margaret Boden, *Artificial Intelligence and Natural Man* (New York: Basic Books, 1977), pp. 434–44; and Pamela McCorduck, *Machines Who Think* (San Francisco: W. H. Freeman, 1979), pp. 180–205.

34. Boden, *Artificial Intelligence,* p. 435–36.

35. McCorduck, *Machines Who Think,* pp. 198–99.

36. See ibid., p. 205; and Boden, *Artificial Intelligence,* pp. 437–41.

37. John R. Searle, "Minds, Brains, and Programs," reprinted in Douglas Hofstadter and Daniel C. Dennett, *The Mind's I* (New York: Basic Books, 1981), pp. 353–73; and John R. Searle, *Minds, Brains, and Science* (Cambridge: Harvard University Press, 1986).

38. Searle, "Minds, Brains, and Programs," p. 358.

39. See Frank Rose, *Into the Heart of the Mind: An American Quest for Artificial Intelligence* (New York: Harper and Row, 1984), pp. 165–69.

40. Putnam, "Robots: Machines or Artificially Created Life?", p. 691.

41. Ibid.
42. Joseph Weizenbaum, *Computer Power and Human Reason* (San Francisco: W. H. Freeman, 1976), p. x.
43. Ibid., pp. 269–70.
44. Sherry Turkle, *The Second Self: Computers and the Human Spirit* (New York: Simon & Schuster, 1984), pp. 265–68.
45. See Marvin Minsky, *The Society of Mind* (New York: Simon & Schuster, 1985).
46. Turkle, *Second Self*, pp. 294, 296.
47. Sherry Turkle, *Psychoanalytic Politics: Freud's French Revolution* (New York: Basic Books, 1978), p. 103.
48. See Gilles Deleuze and Félix Guattari, *Anti-Oedipus: Capitalism and Schizophrenia*, trans. Robert Hurley, Mark Seem, and Helen R. Land (New York: Viking, 1977), Ch. 1.
49. M. Mitchell Waldrop, "Flying the Electric Skies," *Science* 244 (June 30, 1989):1532–34.
50. For a discussion of the debate, see June Goodfield, *Playing God: Genetic Engineering and the Manipulation of Life* (New York: Harper and Row, 1977); Jeremy Cherfas, *Man-Made Life* (New York: Pantheon, 1982); Sheldon Krimsky, *Genetic Alchemy: The Social History of the Recombinant DNA Controversy* (Cambridge, Mass.: MIT Press, 1983); and James D. Watson and John Tooze, *The DNA Story: A Documentary History of Gene Cloning* (San Francisco: W. H. Freeman, 1981).
51. The letter is reprinted in Watson and Tooze, *The DNA Story*, p. 6.
52. The "Berg letter" is reprinted in ibid., p. 11.
53. Robert Sinsheimer, "Troubled Dawn for Genetic Engineering," *New Scientist* 68 (October 16, 1975):148, reprinted in Watson and Tooze, *The DNA Story*, pp. 52–55.
54. Ibid.
55. See Cherfas, *Man-Made Life*, p. 131.
56. See Paul Berg, David Baltimore, Sydney Brenner, Richard Roblin, III, and Maxine Singer, "Asilomar Conference on Recombinant DNA Molecules," *Science* 188 (June 6, 1975):99, reprinted in Watson and Tooze, *The DNA Story*, pp. 44–47.
57. Letter from Waclaw Szybalski to Paul Berg (October 8, 1974), reprinted in Watson and Tooze, *The DNA Story*, p. 16.
58. Cherfas, *Man-Made Life*, p. 131.
59. See the report by Nicholas Wade, "Recombinant DNA: NIH Sets Strict Rules to Launch New Technology," *Science* 190 (December 19, 1975):1175, reprinted in Watson and Tooze, *The DNA Story*, pp. 70–72.
60. Roy Curtiss, III, "Biological Containment: The Construction of Safer *E. coli* Strains," in David Jackson and Stephen Stich, eds., *The Recombinant DNA Debate* (Englewood Cliffs, N.J.: Prentice-Hall, 1979), p. 69.
61. Ibid., pp. 70–79.
62. Krimsky, *Genetic Alchemy*, pp. 213–32.
63. For a comparison of the guidelines, see the excerpt from the *Federal Register* reprinted in Watson and Tooze, *The DNA Story*, pp. 349–56.
64. See Robert L. Sinsheimer, "Two Lectures on Recombinant DNA Research," in Jackson and Stich, *The Recombinant DNA Debate*, pp. 85–98.
65. See Robert L. Sinsheimer, "The Presumptions of Science," in Gerald Holton and Robert Morison, eds., *Limits of Scientific Inquiry* (New York: W. W. Norton, 1979), pp. 23–36.
66. See Bernard D. Davis, "The Recombinant DNA Scenarios: Andromeda Strain, Chimera, and Golem," *The American Scientist* 65 (1977):547–55; and Bernard D.

Davis, "Evolution, Epidemiology, and Recombinant DNA," in Jackson and Stich, *The Recombinant DNA Debate,* pp. 137–54.

67. Davis, "The Recombinant DNA Scenarios," p. 550.

68. Davis, "Evolution, Epidemiology, and Recombinant DNA," p. 151.

69. See the letter from Stanfield Rogers to James Watson (March 20, 1975), reprinted in Watson and Tooze, *The DNA Story,* p. 43.

70. See "Open Letter from Science for the People to the Asilomar Conference," reprinted in Watson and Tooze, *The DNA Story,* p. 49.

71. Lewis Thomas, "The Hazards of Science," *New England Journal of Medicine* 296 (February 10, 1977):324, reprinted in Watson and Tooze, *The DNA Story,* p. 242.

72. For a discussion of this problem, see Paul Ramsey, *The Patient as Person* (New Haven: Yale University Press, 1970); Paul Ramsey, *Ethics at the Edges of Life* (New Haven: Yale University Press,1978); and Richard McCormick, "To Save or Let Die: The Dilemma of Modern Medicine," *Journal of the American Medical Association* 229 (July 8, 1974):172–76.

73. Ad Hoc Committee of the Harvard Medical School to Examine the Definition of Brain Death, "A Definition of Irreversible Coma," *Journal of the American Medical Association* 205 (1968):337.

74. Ramsey, *Patient as Person,* p. 84.

75. For a history of environmental ethics, see Roderick Frazier Nash, *The Rights of Nature: A History of Environmental Ethics* (Madison: The University of Wisconsin Press, 1989). For the work of Aldo Leopold, see *Aldo Leopold, A Sand County Almanac* (New York: Oxford University Press, 1949); Curt Meine, *Aldo Leopold: His Life and Work* (Madison: The University of Wisconsin Press, 1988); and J. Baird Callicott, "The Search for an Environmental Ethic," in Tom Regan, ed., *Matters of Life and Death,* 2d ed.(New York: Random House, 1986), pp. 381–424.

76. Leopold, *Sand County Almanac,* p. 204.

77. Ibid.

78. Ibid., pp. 224–25.

79. See Nash, *Rights of Nature,* pp. 64, 77.

80. Christopher D. Stone, *Should Trees Have Standing?—Toward Legal Rights for Natural Objects* (Los Altos, California: William Kaufmann,1974), p. 5. This work first appeared as an article in the *Southern California Law Review* in 1972.

81. See John J. Costonis, "The Chicago Plan: Incentive Zoning and the Preservation of Urban Landmarks," *Harvard Law Review* 85 (1972):574; John J. Costonis, "Do Buildings Have Rights?" *Student Lawyer* (December 1975):13–17; and M. P. and N. H. Golding, "Why Preserve Landmarks?—A Preliminary Inquiry," in K. E. Goodpaster and K. M. Sayre, eds., *Ethics and Problems of the 21st Century* (Notre Dame, Indiana: Notre Dame University Press, 1979), pp. 175–90.

82. Nathaniel Hawthorne, "The Artist of the Beautiful," in *Moses from an Old Manse, The Centenary Edition of the Works of Nathaniel Hawthorne,* vol. 10 (Columbus: The Ohio State University Press, 1974), pp. 447–75. The story was originally published in 1844.

83. Ibid., p. 471.

84. Richard Brautigan, "All Watched Over by Machines of Loving Grace," in *The Pill Versus the Springhill Mine Disaster: Selected Poems, 1957–1968* (New York: Dell Publishing, 1968), p. 1.

85. Leopold, *Sand County Almanac,* p. 226.

Index